I0112562

BECOMING MARTIAN

The MIT Press's publishing mission benefits from the generosity of our donors, including Sally Yu.

BECOMING MARTIAN

HOW LIVING IN SPACE WILL CHANGE OUR BODIES AND MINDS

SCOTT SOLOMON

FOREWORD BY SCOTT KELLY

THE MIT PRESS CAMBRIDGE, MASSACHUSETTS LONDON, ENGLAND

The MIT Press
Massachusetts Institute of Technology
77 Massachusetts Avenue
Cambridge, MA 02139
mitpress.mit.edu

The MIT Press would like to thank the anonymous peer reviewers who provided comments on drafts of this book. The generous work of academic experts is essential for establishing the authority and quality of our publications. We acknowledge with gratitude the contributions of these otherwise uncredited readers.

This book was set in ITC Stone and Avenir by New Best-set Typesetters Ltd. Printed and bound in the United States of America.

Library of Congress Cataloging-in-Publication Data is available.

ISBN: 978-0-262-05151-4

10 9 8 7 6 5 4 3 2 1

EU Authorised Representative: Easy Access System Europe, Mustamäe tee 50, 10621 Tallinn, Estonia | Email: gpsr.requests@easproject.com

To Nyala, Nicholas, and Thomas

Earth is the cradle of humanity, but one cannot live in a cradle forever.
—Russian space pioneer Konstantin Tsiolkovsky

Like Darwin, we have set sail upon an ocean—the cosmic sea of the Universe. There can be no turning back. To do so could well prove to be a guarantee of extinction. When a nation, or a race or a planet turns its back on the future, to concentrate on the present, it cannot see what lies ahead. It can neither plan nor prepare for the future, and thus discards the vital opportunity for determining its evolutionary heritage and perhaps its very survival.
—former NASA administrator James Fletcher

The future of the human race itself may eventually depend on one question: "Can we have sex in space?"
—theoretical physicist Dr. Michio Kaku

CONTENTS

FOREWORD

Leaving planet Earth is something you never forget. My first experience came aboard the space shuttle *Discovery* in 1999. I was strapped into my seat with six crewmates as seven million pounds of thrust launched us off the pad from the Kennedy Space Center in Florida. The vibrations felt like riding a freight train that had jumped the tracks but continued to accelerate. As the g-forces built, less than a minute later, we were traveling faster than the speed of sound. As we continued to accelerate to upward of three g's, even breathing became a challenge.

Then, suddenly, the main engines cut off. At 17,500 miles per hour, we were now orbiting the Earth. Yet, inside the shuttle, it felt like we weren't moving at all. As a Navy test pilot, I had flown jets exceeding Mach 2, but this was something entirely different. We were floating. The sensation was strange but exhilarating, something I'd only briefly experienced during parabolic training flights. I looked out the window and saw a view that my brain struggled to process: a glowing blue and orange curve, the Sun rising over Earth. It was breathtaking—the most beautiful sight I'd ever seen. Like someone painted the most brilliant blue paint on a piece of glass. I was in awe.

By our second day in space, we reached Hubble in a high orbit—150 miles above the International Space Station's orbit, which years later would become a second home to me. Our mission lasted eight days, during which we repaired the telescope and installed new scientific instruments. We completed our tasks successfully and celebrated Christmas on the seventh day—a surreal but unforgettable experience—the only Christmas in space for a space shuttle crew. Adapting to weightlessness took time, but I was fortunate not to suffer from space motion sickness, at least not in any severe way, which affects many astronauts. My more experienced crewmates (I was the only rookie) taught me how to navigate life in microgravity: how to move in a controlled and deliberate manner, how to work, even how to brush my teeth and use the restroom. Over time, these things became second nature, and I adjusted to the reality that there's no such thing as "up" or "down" in space.

I was lucky to return to space three more times: once aboard space shuttle *Endeavour* and twice on Russian Soyuz rockets launched from Kazakhstan in Central Asia. My final mission was the longest: 340 consecutive days on board the International Space Station. Alongside my Russian colleague Mikhail Kornienko and a rotating crew of thirteen other astronauts and cosmonauts, we conducted scores of scientific experiments over the course of the year. One primary focus was understanding how the human body adapts to extended periods in space—essential knowledge if we aim to send people to Mars.

The studies were comprehensive. Researchers collected samples of my blood, skin, saliva, urine, and feces before, during, and after the mission. Onboard, I handled these tasks myself, storing the samples for return to Earth. I swabbed my skin for microbial studies, performed ultrasounds on my own anatomy, and monitored everything from sleep patterns to reaction times and mood. My identical twin brother, Mark, served as a control back on Earth, enabling researchers to compare how space uniquely altered my body.

Long-term spaceflight takes a toll, physically and psychologically. Two hours of daily exercise are required to stave off muscle and bone loss, yet recovery on Earth takes months. Gravity feels foreign after such a prolonged absence. Headaches, dizziness, nausea, and skin rashes are common. My vision worsened during my first extended mission, and scans revealed that small folds had developed behind my retina. Living in confined quarters with just a few crewmates also challenges mental resilience.

But the sacrifices are worth it. I'd dreamed of being an astronaut since reading *The Right Stuff* by Tom Wolfe in college. Until then, I had little direction in life. Watching Neil Armstrong and Buzz Aldrin walk on the Moon as a child had been thrilling but also terrifying; I'd even had recurring nightmares about rockets exploding. Wolfe's book changed my perspective. It captured the courage and determination of test pilots pushing human limits, inspiring me to pursue that path—first as a Navy fighter pilot, then a test pilot, and later as a NASA astronaut.

My year in space offered a glimpse into what a journey to Mars might entail. Living in space is nothing like a brief visit, certainly not a vacation; it's more akin to moving there. Each mission, my body adapted a little faster, but the physical changes were profound. Mars travelers will face similar challenges, and if humans are to live there, we need to understand how its conditions will affect us. One question I often ponder is whether someone born on Mars could ever visit Earth. Returning to Earth's gravity after my missions was among the most physically demanding experiences of my life. Could a person who's never experienced Earth's gravity handle it?

While Mars represents a bold frontier, nothing makes you appreciate Earth quite like leaving it. The wind on your face, the smell of grass, the feeling of water—these simple pleasures become treasures you deeply miss in space. I don't know how Mars settlers will feel about Earth, but I hope they'll visit and experience the beauty of this extraordinary planet we call home.

—Former NASA astronaut Scott Kelly

INTRODUCTION

There are few things that can put life in perspective quite as well as a canoe trip down the Brazos River. There's a particular section of it, not far from where I live, where an otherwise enjoyable paddling excursion can prompt a person to contemplate humanity's past, present, and future. The Brazos winds its way south and east from the panhandle region of Texas down toward the Gulf of Mexico. The lower section is mostly a muddy affair, the kind of place that could really ruin a pair of shoes. But there is one spot, between Waco and College Station, that is unlike the rest. Here, when the river runs low, there's a small set of rapids where the riverbed changes from soft mud to hard sandstone. If you pull your canoe up along these rocks and take a close look, as I did one day in late fall, you will see something remarkable.

The sandstone appeared to have ripples—tiny ridges that run perpendicular to the river's flow. Luckily, I had come with a geologist, Cin-Ty Lee, who helped me understand what I was looking at. "These are tsunami deposits," Cin-Ty explained. "The wave ripped right through here."

Cin-Ty and I were standing some 175 miles from the coast. But at the end of the Cretaceous Period, 66 million years ago, this area was in the middle of a shallow sea. The wave that caused the deposits we were examining was estimated to have been about 100 feet high. I looked up, trying to imagine a wave that size. I shuddered at the thought.

The site was discovered by geologists in the 1980s, and the recognition that the ripples in the sandstone were formed by a tsunami helped to provide additional evidence for the theory that an asteroid impact was to blame for the mass extinction event that killed the dinosaurs.[1] The idea was controversial at first, in part because it seemed like an impact large enough to cause death on such a massive scale would have left quite a big crater. Yet no one had found a crater large enough or of the right age to have been the smoking gun needed to prove the theory.

But those ripples on the Brazos River bed were a clue. They were from the right time, precisely at the end of the Cretaceous, when the mass extinctions began. And not only did they show that a massive wave had passed this way at that time; they pointed to the direction that the wave had come from: somewhere between the current Texas coast and the Caribbean.

Sure enough, the signature of a crater buried underground was soon found on Mexico's Yucatán Peninsula near the village of Chicxulub. The crater turned out to be 112 miles across, extending far into the Gulf of Mexico. Core samples showed that it formed at the same time that dinosaurs and 80 percent of the other species alive at the time suddenly disappeared. Bingo.

It's difficult enough to imagine a 100-foot-tall tsunami, much less an approaching asteroid larger than Mount Everest. The impact is thought to have been more powerful than 100 million hydrogen bombs detonating simultaneously. Debris was sent so high in the atmosphere that it traveled around the world, covering the planet in a dark curtain of particles that blocked the sunlight for years. Plants struggled to get enough light. When they died, so too did the many animals that ate them. The tsunami would have been destructive enough, but the cascading wave of death triggered by the years of dark and cold that followed was even more devastating.

It's hard not to wonder if such a calamity could happen again. The reality is that, eventually, it will. Meteors strike the surface of the Earth all the time. Most are small, and cause little to no damage. But larger ones come from time to time. Estimates are that meteors large enough to cause some sort of global impact strike the Earth about every 100,000 years. Those large enough to cause a mass extinction come around less often, but exactly how rare they are is not known. Still, it's not a matter of *if* it will happen, so much as *when*.

On the other hand, seeing the physical evidence of the tsunami, and actually touching it with my hands, I couldn't help but think about the ways in which that single event altered the course of history. If the dinosaurs had not gone extinct, mammals might never have become the world's dominant land animals. That means primates might not have come into existence. And, without primates, there would not be any humans. We owe our existence to that catastrophic disaster.

And if another asteroid that size should ever come our way, it could very well determine our future.

*

Thinking about how life has changed through time, including who we humans are and where we come from, has always fascinated me. As a kid, seeing fossils of dinosaurs and other animals in natural history museums filled me with awe and wonder. I spent hours digging in my backyard, searching in vain for the remains of some extinct creature. In college, I discovered there was an entire field of study dedicated to tracing the connections between all living things and figuring out what how they change over time—the field of evolutionary biology. And in an anthropology class I learned how these tools have been used to piece together our own origins as humans. I became intrigued by the notion that we could reconstruct the steps that led a group of apes living a simple existence on the African plains to become a species capable of building cities with towering skyscrapers, devices to communicate instantly across the world, and rockets that would carry them to the Moon.

As a graduate student, my interests shifted toward understanding the evolutionary origins of complex societies of a different sort—those of insects. My PhD research brought me to the jungles of South America, where I excavated enormous nests containing millions of leafcutter ants. Their ability to grow their own food, a type of fungus they cultivate inside their nests, is eerily reminiscent of human agriculture. Collecting hundreds of samples of these ants and the fungi growing in their subterranean gardens led to a remarkable discovery. Our genetic analyses showed that the timing for when the ancestors of leafcutter ants began farming fungi coincided remarkably well with the asteroid impact and the mass extinction event it triggered at the end of the Cretaceous period.[2] While we can't be sure what prompted those first ants to start growing fungi, it seems plausible that the two events are connected. In a dark world, surrounded by the decaying remains of the animals and plants that succumbed to the impact and its aftermath, fungi would have been one of the few reliable sources of food.

When I got my first faculty position in 2009 at Rice University in Houston, Texas, I found myself thinking once again about human evolution. But now, rather than thinking just about our origins, I began thinking about what's happening today. Are humans still evolving? I put this question to the students in the introductory biology course I was teaching, hoping it would encourage them to do some deep thinking. It did, and the ensuing discussion prompted me to do a deeper dive into what we know about human evolution in modern times and how our evolution may proceed in the future. The result was my first book, published in 2016, entitled *Future Humans: Inside the Science of Our Continuing Evolution.*[3]

My main conclusion was that, yes, humans are still evolving. But modernization has changed the specific ways in which evolution is happening. Charles Darwin's theory was that evolution happens by natural selection, and that certainly is true for humans as it is for all species. But other evolutionary mechanisms are at work, too, including some that were not yet known during Darwin's lifetime. Toward the end of the book, I considered the possible ways that our evolution might proceed in the distant future. Our ultimate fate as a species, I surmised, comes down to three possibilities.

One option is that our species remains as it is indefinitely. So far, in the 3.7-billion-year history of life on Earth this has not happened to a single species, and I argued that it wouldn't happen to us, either. All species change, some faster than others, but there is no species alive today that has not undergone changes throughout its existence. On the other hand, we could become extinct, as the vast majority of species that preceded us have done. There are an uncomfortably high number of plausible ways this could happen, including another giant asteroid impact, a supervolcano eruption, nuclear war, catastrophic climate change, the spread of a devastating pandemic, or our Sun exploding in a supernova. Clearly, we would like to do everything we can to avoid that fate.

The last possibility is that we will evolve into another species—a new type of human. I considered the ways this could happen and argued that as long as things remain more or less as they are today—with a large global population of humans, interconnected yet spread out around the world—that this is unlikely. That is, unless something major changes. As an example of the kind of game-changing development that could, perhaps, prompt humanity to split and evolve into one or more new species I suggested a scenario in which some humans leave Earth and create settlements somewhere in outer space.

It was an intriguing hypothetical thought, but at the time it seemed like little more than a science fiction fantasy. I met a few people who were giving the idea of space settlement more serious thought, but I figured this had a lot to do with the fact that I was a science geek who happened to live in Houston, aka Space City. After all, we've got NASA's Johnson Space Center, the home of NASA's Mission Control and the place where astronauts have trained since before the Apollo era. Rice University has a long history of research on space science, with close ties to NASA. I sat in on a few talks that discussed space settlement, but it didn't seem to me to be the kind of thing many people were actually pursuing in a serious way.

My perspective changed, though, soon after my book came out. I recall being on a flight to New York for one of the first events in my book tour and reading an article in an in-flight magazine arguing not only that we should build settlements on Mars, but that we should not stop there. We need to keep spreading out deeper into space, to other planets and even other solar systems, the author suggested. Perhaps the concept of space settlement was more mainstream than I had realized. Indeed, as the publicity for my book rolled on, I found myself answering more and more questions about the idea that leaving Earth would cause us to evolve in new ways. I was invited to give talks on the topic and to write articles expanding on the ideas I had briefly sketched out in the book. It seemed like everyone wanted to talk about what will happen to us in space.

The more I looked into it, the more I realized that people weren't just talking about a human migration into space—there were some who were actively pursuing it. The most conspicuous effort began emerging a few years later from a strip of tidal flats at the southern tip of Texas. There, at a place known as Boca Chica, SpaceX began building a production and launch facility in 2014. Their first experimental rockets started launching five years later from the facility, which became known as Starbase. And in April 2023, SpaceX announced that they would conduct the first flight test of a revolutionary new spacecraft known as Starship. NASA had contracted a Starship for its Artemis III mission to the Moon, but SpaceX CEO Elon Musk said that Starship was designed to do much more—to carry people and cargo to build a city on Mars. While SpaceX had been remarkably successful in designing reusable rockets that launch satellites and regularly transport people and supplies to the International Space Station, nothing as large and ambitious as Starship had ever been attempted by them or anyone else.

I decided I needed to see this for myself.

*

A public road passes right through Starbase so, incredibly, anyone can drive right up to the launch pad. And many do. When I arrived just one day before the launch window opened, the place was crawling with people. They ranged from middle-aged men wearing shirts that read "Occupy Mars" to families with children wearing astronaut costumes. Cars, trucks, and camper vans were parked all along the road on the side opposite from the launch pad. Beside the vehicles people were standing or sitting in folding chairs, looking through binoculars and snapping photos. It felt like a mix between a music festival and a ComicCon convention.

Just a few hundred feet from the beach, the launch tower rose up from the dunes. Sitting on the launch pad was an enormous monstrosity. The lower portion, a metallic cylinder called the Super Heavy booster, was gleaming in the bright sunlight. The top portion was the second stage, covered in black on one side with two pairs of fins that stuck out. The second stage, called Starship, was designed to separate from the booster after launch and continue on to destinations in deep space. Stacked together, as they were, the booster and second stage made for the largest rocket ever built.

On a low concrete wall beside the road someone had painted the words "To Mars & Beyond . . ." SpaceX employees in hard hats busily moved about. Between the tourist vehicles, large trucks carrying liquid helium and nitrogen slowly pulled up to the launch facility. Large plumes of white gas billowed out as they transferred their contents to a series of four interconnected cylindrical white towers just off the road. Periodically, a loud hissing sound came from the rocket assembly just behind the towers.

The launch window opened at 7:00 a.m. the following morning. The public road to the launch facility had closed that morning to ensure public safety, so I joined the crowd of people on the beach at the southern tip of South Padre Island with a clear view of the launch from five miles away. As the Sun rose over the gulf, thousands of people set up folding chairs and tripods in the dunes and along the sea wall. Just across the channel to our south, the rising Sun cast a red glow off the reflective surface of the Super Heavy booster and the darker Starship stacked on top. Despite near perfect weather conditions, with less than fifteen minutes left in the countdown a collective sigh rippled through the crowd. The launch was being "scrubbed" due to an issue with pressurization in the booster. Chairs folded, cameras were loaded into backpacks, and the disappointed crowd scattered from the beach like so many crabs.

The next launch attempt would be three days later. Sadly, I couldn't stay. From my home in Houston, I watched the countdown pick up again along with 72,000 other viewers via a live stream on YouTube called "Everyday Astronaut." Combined with other channels, there were nearly a million people tuning in for the event. There was a palpable excitement as the countdown reached the single digits. Suddenly, a blindingly bright light emerged from the bottom of the rocket, followed by thick plumes of gray smoke. I could see on the screen that the clouds completely engulfed the road where I had stood just three days earlier. For a moment, the rocket just sat there on the pad. Then, slowly at first, it began to ascend. The color of the rocket's flame turned from a yellow to a brilliant silver tinged with purple. A graphic at the bottom of the screen showed that three of the booster's thirty-three engines had failed to fire. Soon, a few more blinked out. But that didn't stop the enormous rocket from climbing ever more rapidly into the sky.[4]

Then, three minutes after launch, the rocket suddenly exploded. SpaceX officials called it a RUD—short for "rapid unscheduled disassembly," and later explained that the vehicle's automatic detonation system had been activated after it veered off its planned course. While many news headlines focused on the fiery end to the test launch, others saw it as a useful step forward. "Congrats @SpaceX team on an exciting test launch of Starship! Learned a lot for next test launch in a few months," Elon Musk tweeted.

In his article on the day of the launch for the online technology news outlet Ars Technica, space journalist Eric Berger discussed the explosion, noting that SpaceX doesn't wait until their rockets are flawless before testing them. "For those who know a bit more about the launch industry and the iterative design methodology, getting the Super Heavy rocket and Starship upper stage off the launch pad was a huge success," he wrote. In addition to the explosion, Berger noted that the launch pad had sustained a fair amount of damage. Still, he was optimistic that Starship would eventually be able to fly as intended. "When SpaceX irons out all of these issues, we'll be left with the world's largest fully reusable rocket. This will forever change humanity's relationship with the cosmos . . . we are seeing the transformation of life off Earth," he wrote.[5]

A fleet of fully operational Starships would be a total game changer for access to space. It was designed to carry up to 150 metric tons while being fully reusable, meaning that the cost to launch that weight will be substantially lower than with a disposable rocket. Berger estimated that a launch of the reusable Starship could cost as little as $30 million. Others have speculated that the cost could go down to $10 million. Musk, who has a reputation for overly optimistic projections, stated it could drop to just $2 million per launch. Even with the upper estimates, having such an inexpensive way to send heavy loads to space would make it economically possible for the first time in history for humans to build a permanent base on another celestial body like the Moon or Mars.

Indeed, that is what Starship was designed to do. Musk has repeatedly made it clear that SpaceX's long-term goal is to create a permanent human presence on the Red Planet. In February 2024, he posted on his social media platform, X,[6] "We are mapping out a game plan to get a million people to Mars. Civilization only passes the single-planet Great Filter when Mars can survive even if Earth supply ships stop coming." What's more, he plans to begin creating settlements on Mars very soon. In another X post, in May 2024, Musk laid out his timeline for Mars:[7] "Maybe a city in 20 years, but for sure in 30."

If these plans to create human settlements in space are successful, it could help humanity avoid the catastrophic destruction of another massive meteor impact, or any of the other threats that our species faces if we are restricted to just one planet. As it stands, we have all our eggs in one cosmic basket, so to speak. Indeed, keeping our species alive is a major motivating factor for many advocates of building human settlements on other worlds. But some people point to other reasons for leaving Earth, from the potential for profitable enterprises like space mining or manufacturing to easing the ecological or population burden on our home planet, or simply because it would make for an exciting adventure.[8]

Jeff Bezos, founder of Amazon and the rocket company Blue Origin, sees moving to space as an opportunity for human potential. "I would love to see a trillion humans living in the solar system," Bezos said in a podcast interview in 2023.[9] "If we had a trillion humans, we would have, at any given time, 1,000 Mozarts and 1,000 Einsteins . . . our solar system would be full of life and intelligence and energy. And we can easily support a civilization that large with all of the resources in the solar system."

Like Musk, Bezos has put his money where his mouth is. Bezos reportedly invested some $10 to $20 billion of his funds into Blue Origin, which has the goal of facilitating human migration into space.[10] While some have argued that Bezos and Musk are idealistic, they are two of the wealthiest people on Earth, and both have poured significant amounts of their own resources into developing technologies to help achieve their visions for humanity's future in space. That doesn't make them right, but it at least suggests they are confident.

✷

While the billionaires with their own rocket companies tend to attract a lot of attention, they are by no means the only ones actively working toward a future that includes people living in space. To meet more of these folks, I traveled to Washington, DC, a few weeks after the first Starship launch for a conference called the Humans to Mars Summit. It was held at the stately National Academy of Sciences building, which feels like a temple to science. Hanging on the walls below the ornate arches and domes were portraits of scientists. There was a bust of Albert Einstein and another of Charles Darwin. Its elegant library is filled with books about physics, chemistry, and medicine. For a science geek like me, it felt like a holy place. The talks took place in a futuristic-looking auditorium with large white triangular panels forming the walls and ceiling illuminated by red lights that helped give the proceedings a space-age feel.

One of the first speakers was NASA associate administrator Robert Cabana. He explained that NASA's current goal is to send humans back to the Moon as a key step in the agency's broader "Moon to Mars Architecture." They are planning to build a permanent base on the Moon where astronauts will live and work for months at a time. The idea is to practice on the Moon for what will be needed to have people on Mars for extended stays.

Cabana was asked why we should go to Mars. After all, the questioner noted cheekily, Elton John warned us in his song "Rocket Man" that "Mars ain't the kind of place to raise your kids—in fact it's cold as hell." The crowd chuckled. While Cabana didn't debate the finer points of Martian meteorology or childcare, he was sanguine about the motivation for going. "We go to explore, to learn, to expand our knowledge of this vast universe that we live in," he responded. "It's so important . . . we need to send humans. We need to establish that presence in our solar system beyond our home planet."

In addition to many other NASA officials, the speakers also included representatives from other government space agencies, including the European Space Agency (ESA), and the Japan Aerospace Exploration Agency (JAXA), as well as private space organizations. One after another, they stepped on stage and spoke about their visions for humanity's future in space.[11]

"I'm here to tell you that we are in fact going to Mars," said Tory Bruno, president and CEO of United Launch Alliance, a joint venture between commercial aerospace giants Boeing and Lockheed Martin. "We are going to send people there . . . it is our human destiny to extend our presence beyond this planet," he said.

I was astonished to see how extensive and serious the plans for human space exploration had become. Listening to all these talks by industry professionals and speaking with them during breaks, it became clear to me that having humans in space was not merely a possibility for the future—it is happening. I came to the summit in DC naively expecting to hear from people fantasizing about a vision based more on science fiction than actual science. Instead, I learned that some of the world's leading engineers, scientists, and business professionals genuinely think there will soon be people setting foot on Mars—perhaps within their lifetimes—and that having people living there could soon follow.

But so much of the focus seemed to be on how to get people to Mars and how to keep them alive. I gathered that this was an improvement over earlier times when the focus of much of the space industry was almost entirely on engineering and tended to overlook humans or treat them as an afterthought. Indeed, the organizer of the Humans to Mars Summit, Chris Carberry, told me that he and his co-organizers pride themselves on including much more programming related to "human elements" than do many other space exploration conferences.

Still, I couldn't help but notice that there wasn't much discussion about the long-term consequences if these efforts are successful. If we create permanent human settlements on Mars, what should we expect to happen to its inhabitants over time? Would there be evolutionary changes in later generations? Would their bodies adapt? Could people born on Mars come back to Earth? Would new species of humans eventually emerge?

This book chronicles my attempt to investigate these questions. I wanted to find out what we actually know about how space affects the human body and mind—and how we know it. I wanted to see if our knowledge about evolution and about our species' origins could help us to predict our possible future among the stars. This quest turned into a fascinating journey. I sat down with astronauts to learn about their personal experiences in space and visited research labs so secure they need iris scanners. Along the way I met a motley crew of fascinating people, including nuclear physicists, microbiologists, philosophers, rocket engineers, historians, geologists, medical doctors, YouTubers, entrepreneurs, anthropologists, geneticists, sociologists, and many others. I've been constantly impressed by not only how smart these people are, but also how welcoming and friendly they have been. They invited me to their laboratories, launch sites, training facilities, conferences, workshops, lecture halls, favorite coffee spots and restaurants, and even their homes. For a kid who grew up with a fascination for science and discovery, getting to know these amazing people, learning about what they do, and watching them do it has been a thrill. This book tells some of their stories.

But at the heart of this book is also a fundamental question. We know we'll face threats in the future, whether from deadly asteroids or ecological meltdown as a consequence of our own actions. Some say that leaving Earth is the only way to ensure our long-term survival. But given what we're learning about how migrating into space would change us, should we do it? And if we decide the answer is yes, when should we begin? What sacrifices should we be willing to make in order to ensure these efforts are successful? In short, how do we move forward if it turns out that moving to Mars means becoming Martian?

To address these questions, we'll begin by exploring the most likely site of our first space settlement, Mars, and consider how we know what it's like there. Chapters 2 and 3 trace our emerging understanding of how the conditions in space affect the human body. Chapter 4 examines one of the biggest questions in our efforts to become a multiplanetary species: the question of whether we can reproduce in space. Chapter 5 explores the psychological changes that could come from leaving Earth and living in space. Chapters 6 and 7 consider what evolutionary changes could occur among humans and any other organisms we bring with us. Finally, chapter 8 asks whether we should let evolution run its course or take matters into our own hands. Then, I wrap it up by returning to the questions that prompted me to begin this journey in the first place—given what we know, should humans leave Earth? I'll withhold my opinions for now so you can come to your own conclusions as you join me in this quest. Along the way I will introduce you to the people who are doing all this work and take you to many of the places where it's happening.

Put on your space helmet and strap into your rocket seat—we're going for a ride.

1

THE RED PLANET

It was Sol 1110 on Mars, and a team of scientists was working to figure out the best way to navigate a particularly rugged stretch of terrain. The destination could be seen in the distance, an outcrop at the edge of a narrow canyon known as Bright Angel. It was too far to reach in a day—or, as the days are known on Mars, a sol—but it might be possible to get there in a week or so if things went well. Then again, the going had been slow for the last several sols. Yesterday the progress had only been about twenty-four meters.

I was sitting in a conference room on the Rice University campus next to two of the scientists on the team, Kirsten Siebach and Ellie Moreland. On the screen in front us we could see the barren, rocky terrain and the current position of the rover. Despite the fact that we were some 310 million kilometers away, the images were clear enough to see every crack and crevice on the rocks surrounding the rover as well as the reddish-orange sand that filled the space between the rocks and the pale, yellow sky above the horizon. At first glance, the landscape looked as if it could have been from a desert valley someplace on Earth. I had to remind myself that we were actually looking at a scene from another planet, getting as close as you can get to a live view of another world. The idea gave me goosebumps.

The images had been sent just hours earlier from Jezero Crater on the surface of Mars to one of several satellites orbiting the Red Planet. From there they were relayed to one of three stations on Earth located in the United States, Spain, and Australia, known collectively as the Deep Space Network. That whole process took about twenty minutes. The science team was now busy looking through the images and data and using them to revise its plans for the sol. Once the scientists came to an agreement—a process that involves multiple, increasingly detailed meetings spanning much of the day—they would send instructions back through the Deep Space Network to the rover, which would execute them the following sol.

The first meeting, called Plan Development, was about making any changes to the plan for Sol 1111 based on new data just received from the rover. The participants on the call included scientists and engineers from around the world, each of whom had a specific role and responsibility. The Tactical Science Lead was Jeff Johnson. He had a very calm, soothing voice like the kind you might hear on a guided meditation app. "I put in a couple of placeholders just to get the conversation started," Johnson announced to the group. On the screen I could see a few rocks that had been labeled with a five-digit number. "If folks zoom in on that 10987 target they'll see it's one of these rocks that has some of the darker, smoother coating remnants on it," he added.

Ellie Moreland, a graduate student working with Siebach, was sitting across from me in the meeting room in the geology building on the Rice University campus in Houston. She explained that any scientist on the call could place a marker on any point in the landscape surrounding the rover, which the whole team would see on their screens, and then they could make an argument for why that would be an interesting and worthwhile place to collect some type of data. The rover, named Perseverance, was equipped with an extensive toolkit.[1] Each day, the scientists send instructions to the rover telling it to perform a specific set of tasks. But first they need to come to an agreement about what exactly those tasks are, and where exactly each should be done.

"I put in the 10986 and 10989 over on the left side of the rover. I could talk to those," offered Jens Frydenvang, referring to the numbers that now appeared over two of the rocks on our screen.

"Sure," Johnson responded.

"I opted for the sort of bigger blocks we have over on the left side of the rover. I put in an early footprint for the 10989, but I was thinking 10986 and 10989 would be similar," Frydenvang explained.

"Right, yeah. I think that's what I was seeing as well," said Johnson. "10986 is rugged enough that it's probably got a fair amount of soil and dust contamination in the little depressions."

The discussion went on like this for several minutes, with several people weighing in with opinions about the pros and cons of the various points that had been suggested as sampling targets. The biggest constraints on what the rover can do are time and energy. Generally speaking, the rover cannot collect certain kinds of data, like images, while it's driving, so the further it goes each sol, the less time it has for collecting data. The rocky terrain of the region where Perseverance had been navigating the last few sols had limited its movement to short distances, which meant there was a little more time available for science.

"We have some serious power gain," Johnson noted.

"That's good!" Moreland exclaimed with a smile. "Usually, we're very tight on power," she told me.

When the rover travels a shorter distance it uses less power, so the slow progress of the rover the last few sols meant there was more electricity available to run the scientific equipment. Perseverance runs on nuclear power, using the radioactive decay of a 10.6-pound chunk of plutonium to generate electricity to charge the rover's batteries. Unlike some earlier rovers that relied on solar power, the nuclear power station on Perseverance operates day and night and isn't affected by the frequent Martian dust storms. The surplus power available to the rover at the moment meant that the scientists could collect more than the usual data using Perseverance's instruments.

To explain to me what each of the instruments on the rover can do, Siebach brought a model of Perseverance made of LEGO bricks. "LEGO made this one super detailed so it can actually help with the conversation," she said. Using the model, she walked me through each of the major parts of the rover as well as the seven major instruments it contains.

"This is the arm," she said, pointing to a device that was extended out in front of the toy rover. "Usually, the arm is tucked in. Most days we would never even bring the arm out." She explained that the arm has a drill with a variety of drill bits that can be used to scrape off the surface of a rock or to collect a core sample. There is also an instrument at the end of the arm called PIXL, short for Planetary Instrument for X-Ray Lithochemistry. It can analyze a rock by examining it up close using X-rays, allowing it to determine the rock's precise chemical composition.

The rover's arm also includes a cleverly named pair of instruments called SHERLOC and WATSON. SHERLOC uses a laser to scan rocks and analyze their chemical composition. Much like the fictional detective in the Sherlock Holmes stories, WATSON is SHERLOC's trusty sidekick. WATSON takes very high-resolution images of the materials that SHERLOC analyzes, allowing scientists to see particles the size of a single grain of sand.

Next Siebach pointed to a small tower extending up from the LEGO rover's body, called the mast. At the top of the mast are cameras that give the rover a wide, panoramic view of its surroundings. The mast is two meters tall, so the cameras on it give about the same perspective as an adult would have who is standing in the same spot. One of the cameras is Mastcam-Z, which can take wide landscape photos and videos, including in 3D, or can zoom in to take a closer look at distant objects. Another pair of cameras on the mast are collectively called SUPERCAM. The mast even has a microphone for recording sound.[2]

In addition to taking photos, the mast also has a high-powered laser that can be pointed very precisely at a target. As the laser hits a target, like a rock, it rapidly heats up the material in the target, causing it to turn into a plasma. The exact chemical composition of the target can then be determined based on the wavelengths of light the plasma gives off. The mast also has a few extensions sticking out from it, much like the boom extending from the mast on a sailboat.

"These booms are measuring wind, temperature, humidity, that they're now getting to do in high resolution today," Siebach explained.

Toward the back of the rover is an instrument called RIMFAX that can perform ground-penetrating radar scans in order to see what lies below the surface. Last, there's a device inside the rover called MOXIE, short for Mars Oxygen In-Situ Resource Utilization Experiment. The purpose of MOXIE was to test whether it's possible to make breathable oxygen from the raw materials in Mars's atmosphere. This technology would be essential to enable humans to survive on Mars. Whereas the air we breathe on Earth has about 21 percent oxygen, the air on Mars has only 0.13 percent oxygen. Most of the atmosphere on Mars is carbon dioxide, which we can't breathe, but which can theoretically be split into its two constituents—carbon and oxygen.

The tests were successful. On September 6, 2023, NASA announced that MOXIE had already completed its mission, having been run a total of sixteen times.[3] It was able to produce up to twelve grams of oxygen per hour, which was twice the rate the engineers had targeted. What's more, the oxygen was high quality—at least 98 percent pure. That was impressive, given that the device was small, about the size of a microwave oven, and only intended as a technology demonstration. The results suggested that a much bigger version of MOXIE—or a large number of them—could make enough oxygen to fill an enclosed habitat with the 21 percent oxygen that our bodies are used to breathing. What's more, any excess oxygen could be used as fuel for rockets departing from Mars to Earth or elsewhere in the solar system.

MOXIE requires a lot of power to run, so when it's operating the rover can't do much else. Sol 1111 would not be a MOXIE day. The team planned to take photos and conduct chemical analyses of specific rocks along the rover's projected path.

The Mars 2020 Perseverance rover landed on the surface of Mars on February 18, 2021. It had been in transit for six months and nineteen days. It spent the next three and half months doing engineering checks to make sure everything was in working order before it could begin its planned mission, which started on June 1—known thereafter as Sol 1. The rover's primary objectives involve searching for evidence of previous life on Mars[4] and collecting samples of rock and "regolith," essentially Martian dirt. Perseverance had already collected twenty-four samples. That left just fourteen more to achieve the planned thirty-eight samples intended to be returned to Earth in a later mission.

Perseverance was the fifth rover to have successfully operated on Mars. Each of the previous rovers had been sent with slightly different mission objectives, but all of them had the same overarching goal: to better understand Mars, including both its history and its current conditions. Ultimately, the robotic missions are a prelude toward sending humans to Mars.

*

It's hard to say when exactly sending people to Mars became a goal for humanity. Ancient people were well aware of the existence of Mars.[5] Many associated it with death or war, perhaps because of its red color. In China during the Han dynasty, Mars was known as the Sparkling Deluder and was considered a bad omen. Seeing it in the sky was a sign of an impending invasion or drought. The Babylonians in ancient Mesopotamia also watched Mars with trepidation. They, too, interpreted its movement in the sky as an indication of future events. If it started moving in the opposite direction of other "stars" (known today as retrograde motion), like those in the constellation Leo, it meant the king's reign was about to end.

The idea that objects in the sky could influence events on Earth seemed intuitive. After all, the apparent movement of the Sun causes day and night, and its position in the sky influences the seasons. The movement of the Moon causes tides, and it goes through clearly visible cycles that were often tied to the timing of events like harvests and religious ceremonies.

In Europe, the idea that Mars affects human affairs here on Earth, generally in a bad way, persisted into medieval times. But what would happen if a person went to Mars? According to the ancient Greek worldview that formed the basis for much of Western thought, it wasn't actually possible to go to Mars. The heavens were thought to be made of ether, considered a fifth element fundamentally different from the four elements—earth, air, water, and fire—that make up our world.

But that couldn't stop the imagination of some medieval authors. In his book *The Divine Comedy*, Dante Alighieri describes an imaginary journey that famously included a vivid description of his travels through Hell (Inferno), Purgatory, and Paradise. Mars was one of the stops on his travels through Paradise. Dante envisioned Mars as the final resting place for martyrs who died virtuously in the name of Christ. Dante described how, as he approached the planet, the light that makes Mars appear bright red from a distance took the shape of a glowing cross made from the souls of the martyrs.

In the late sixteenth century, Danish astronomer Tycho Brahe made very detailed observations of the movement of Mars. The Danish king had given him a small island near Copenhagen where the astronomer built an observatory. Brahe was precise and meticulous. He took measurements again and again to check their accuracy. He was also, apparently, a bit belligerent. He once fought his cousin in a sword duel over a mathematical formula in which he lost part of his nose. For the rest of his life, Brahe wore a metal prosthetic nose.[6]

Brahe died in 1601, but his detailed notes were carefully studied by his protégé, Johannes Kepler. Kepler had been troubled by the retrograde motion of Mars. Rather than considering it a bad omen to see Mars appearing to move backward, Kepler thought there had to be a mathematical explanation. Using Brahe's detailed notes about the position of Mars, Kepler figured out that Mars isn't orbiting in a perfect circle. Rather, the orbit of Mars—and the other planets, it turned out—could be more accurately described as an ellipse.[7] What's more, Kepler confirmed Copernicus's revolutionary idea—no pun intended—that the planets are orbiting around the Sun, not the Earth. Except, based on Kepler's model, the Sun wasn't at the center of the ellipse; it's off to one side.

Kepler also figured out that the speed of the planets changes as they follow their elliptical orbits. When they're closer to the Sun, they move faster, while they slow down on the part of the orbit that is farther from the Sun. He used these principles to calculate how long it takes Mars to complete one orbit around the Sun—687 days. In other words, a year on Mars is nearly twice as long as a year on Earth. The fact that both Earth and Mars are orbiting the same Sun, but that Earth is traveling on the inside track and covering a much shorter distance, means that Earth "passes" Mars on their respective orbits. Finally, Mars's retrograde motion made mathematical sense. From our perspective on Earth, Mars appears to be moving one direction as the two planets get closer to one another, but then, as Earth passes Mars, Mars appears for a while to be moving in the opposite direction.

Impressively, the observations Brahe had made, which had been so helpful to Kepler, were done without the benefit of a telescope. Once the telescope was invented, in the early seventeenth century, it became possible to make much more detailed observations of celestial bodies. Galileo noted that the Moon has peaks and valleys that look remarkably Earth-like. Dutch astronomer Christian Huygens used a telescope to look at the surface of Mars, where he saw a dark shape in the form of an hourglass that he thought was a sea. Italian astronomer Giovanni Cassini made even more detailed observations of Mars. He, too, saw the dark hourglass shape, which turned out to be a large mountain. Cassini made observations of Mars at different times, day after day. He could see the dark feature move across the surface of Mars from one side to the other. Based on his observations, he calculated the rate at which Mars is rotating. Cassini determined that a day on Mars—a sol—is only slightly longer than a day on Earth, at twenty-four hours and forty minutes.

As telescopes improved, astronomers made increasingly detailed observations of Mars. In 1877, as Mars and Earth achieved their closest positions in their relative orbits, Italian astronomer Giovanni Schiaparelli spent eight months making careful observations of the Red Planet from his observatory in Milan. Mars did not appear red to Schiaparelli, who was colorblind. However, like many colorblind people Schiaparelli had a keen eye for contrasts between light and dark, and this may have helped him in his observations of the faint features on the planet's surface. He could clearly see the differences between the slightly brighter and darker areas on Mars, which had been noted by others as well. He interpreted the dark areas as seas and the brighter ones as land masses, including what he thought were continents and islands. He saw bright white areas at one end of the planet, which he concluded were ice caps at the planet's pole. But he also saw thin lines that stretched across the planet's surface. The lines appeared to him to be long and straight. He referred to these using the Italian word "canali," which could mean channel or stream.[8] But English translations of his writing used the word "canal," which implies an artificially constructed waterway rather than a natural one.

Schiaparelli's work had a profound effect on how people thought about Mars. First, it seemed clear that there was water on Mars. The presence of polar ice caps suggested frozen water. But Schiaparelli also concluded there was liquid water in the form of vast oceans as well as in an extensive network of canals that crisscrossed the planet's continents. Water meant the possibility of life. And liquid water meant that conditions were even more suitable for living things, since temperatures on the surface must be mild.

But, at least to some, Schiaparelli's use of the word "canali" to describe the long waterways suggested that someone had built them. At least that was the conclusion of American astronomer Percival Lowell. A wealthy businessman from a prominent New England family, Lowell had lived in Korea and Japan and published several books for American readers about the Far East based on his travels. When he read about Schiaparelli's observations of Mars, Lowell decided to use his wealth to make additional advances in astronomy. In 1894, he built an observatory in northern Arizona where conditions are ideal for viewing the night sky.

Lowell apparently saw straight lines on Mars, too. In his mind, they were too straight to be natural. He reasoned that they were canals that transported fresh water from the ice caps at the planet's poles to the more temperate regions closer to its equator. This suggested to him that the land masses on Mars were dry places, not unlike the deserts that surrounded his observatory in Arizona. To Lowell, the extensive canal network was an indication of the presence not only of life, but of intelligent, sophisticated beings capable of massive construction projects on a global scale. After all, here on Earth, the first major canals, like the Erie and Suez Canals, had only recently been built. And none of them came anywhere close to crossing the entire planet. Whoever was building these canals on Mars was clearly more advanced than the Earthlings of the late nineteenth century.

*

The notion of intelligent beings living on Mars captured not only Lowell's imagination, but that of the public as well. Authors explored the possibility of alien life in creative and exciting ways, contributing to an emerging literary genre that later became known as science fiction. Among the most influential was H. G. Wells. His popular 1898 book, *War of the Worlds*, was built on the premise that Mars was inhabited by beings who were far more advanced than humans and that the Martians were now coming to invade Earth.[9] Luckily for the Earthlings, whose technology is no match for that of the Martians in Wells's tale, it's our planet's microbes that ultimately defeat the aliens.

Science fiction stories about Mars and Martians exploded in the early twentieth century and drew heavily on what was known about the planet at the time. In Edgar Rice Burroughs's 1917 novel *A Princess of Mars*, the protagonist, John Carter, travels to Mars and finds a desert world with a thin atmosphere.[10] The green- and red-skinned Martians grow crops using an extensive system of irrigation canals. Carter finds he has exceptional strength due to the planet's lower gravity, and he uses his abilities to help save a Martian princess.

Meanwhile, the scientific understanding of the actual conditions on Mars remained heavily influenced by the views of Schiaparelli and Lowell.[11] But advances in spectroscopy, which uses the fact that different materials reflect particular wavelengths of light, were making it possible to analyze the chemical makeup of Mars. Water, rocks, ice, and clouds all have characteristic signatures. Using this approach, in the 1950s American astronomer William Sinton concluded that the wavelengths of light reflecting from Mars are consistent with that of some sort of vegetation. He thought that Mars might have lichens, a combination of algae and fungi living symbiotically that are among the hardiest of all life-forms on Earth. To find out whether Mars was indeed covered by some sort of vegetation, we would need a closer view.

The first spacecraft were sent to Mars in the 1960s. On July 14, 1965, Mariner 4 came within 6,118 miles of Mars as part of its flyby mission. Using its onboard camera, it took twenty-two photos of the planet's surface and transmitted them back to Earth.[12] These would be the first photos of another planet taken from space. But the digital signal that was transmitted from Mariner 4's camera had to be processed by a computer to form an image, and that process took time. The engineers at NASA's Jet Propulsion Laboratory couldn't wait to see the results. They improvised a way to get a glimpse at what Mariner 4's camera had captured before the actual photo was available.

They printed strips of paper with the raw, three-digit numbers from the camera that corresponded to how light or dark a particular pixel in the image was. Using pastels purchased from a nearby art store, the engineers colored in the strips of paper based on the number scale, with yellow representing the brightest areas, orange and red the intermediate areas, and brown representing the darkest. When they were finished, they had made an image of the very edge of Mars. On the bottom right, the dark brown colors depict outer space. A curved red strip represents the edge of the planet, behind which is a mosaic of yellow, orange, and red that shows the clouds in its atmosphere and the contours of its surface. When the actual photo was finally ready, it confirmed that the "paint by number" method had been accurate. But the actual photo was in black and white. By coincidence, the pastels available at the art store happened to contain a color palette quite similar to the actual color of Mars. And so, in some ways, the simple, low-tech approach was a more compelling way to visualize Mars up close.

Once all the images from Mariner 4 had been examined in detail, they told a very different story about Mars. It appeared to be covered in craters, much like the Moon. Since the majority of asteroid impacts happened billions of years ago as the solar system was forming, this suggested that there had not been much in the way of erosion on Mars in a long, long time. And there was no indication of any canals or oceans. There was no sign of recent volcanic activity, either. Mariner 4's magnetometers had not detected any evidence of a magnetic field around Mars, like the kind we have on Earth. As it passed behind Mars, Mariner 4 sent a radio signal to Earth that passed through the very edge of the planet, allowing its atmosphere to be measured. The results showed that the atmospheric pressure was indeed extremely low, between 0.06 and 0.1 psi. By comparison, the atmospheric pressure on Earth is about 14.7 psi at sea level. A temperature probe measured daytime temperatures at −148 degrees Fahrenheit on the surface.

Mars was looking much less like the mild, desert garden envisioned by Schiaparelli and Lowell and more like a hostile, dead planet.

Mariner 4's observations were discouraging, at least for those who hoped to find either evidence of existing life on Mars or a planet that is at least capable of sustaining life. Any attempts to travel to Mars, especially with people, were looking much more challenging. Subsequent flybys largely confirmed the initial findings, while revealing even more detail about the planet. In 1969, Mariner 6 and Mariner 7 got even closer to Mars and took a lot more photos and measurements of its surface.[13] While there were a lot of craters, there didn't appear to be quite as many as the Mariner 4 photos had initially suggested. It didn't look that much like the Moon after all. Analysis of the planet's thin atmosphere indicated that it contained 98 percent carbon dioxide. The data also seemed to suggest that the ice at Mars's polar caps was made largely of frozen carbon dioxide.

Mariner 9 was the first spacecraft to orbit Mars, allowing it to make the most detailed observations yet of the planet. The spacecraft's arrival on Mars in November 1971 coincided with a dust storm on a global scale. The dust made it impossible for Mariner 9's cameras to see any surface details from orbit. Only the tops of the highest mountains were visible.

The storm took months to clear, but when it finally did, Mariner 9 was able to observe the surface of Mars in exquisite detail. In addition to giant mountains there were also enormous canyons. The largest, which became known as Valles Marineris in honor of the Mariner mission that discovered it, stretches 2,500 miles along the planet's equator. That's about the same distance as from New York to San Francisco. It's also deep—up to four miles at the deepest spot. The canyon showed evidence of erosion, suggesting the existence of liquid water at the surface—if not now, then at least at some point in the past.[14] And with liquid water comes the possibility for life—whether past, present, or future.

✷

To learn even more, we had to get down to the surface. The first spacecraft to successfully land on the surface of Mars was Viking 1 on July 20, 1976. A second Viking lander followed, touching down on the surface of Mars 4,000 miles away a month and a half after the first. One of their main missions was to search for evidence of life. The landers conducted three separate experiments, each using a different approach to detect chemical signatures of microorganisms currently inhabiting the Red Planet.

One of the experiments involved looking for evidence of metabolism, one of the signatures of all life as we know it. If the metabolisms of organisms on Mars are anything like those of organisms on Earth, they will take in some type of carbon-based food as an energy source, digest it, and release some of the carbon as part of their digestive process. American engineer Gilbert Levin designed a clever test to see if this was happening on Mars. Some food containing radioactively labeled carbon was added to a sample of Martian soil. If there were living organisms in the soil, Levin hypothesized, they would take in the labeled carbon and release it again as part of their metabolism. Intriguingly, the experiments on both landers did indicate that labeled carbon had been released.

Had Viking just detected evidence for life on Mars?

Levin believed that it had.[15] While the labeled carbon could have been generated by natural chemical processes that do not involve living organisms, Levin reasoned that the second part of his experiment would be able to determine if that was the case. After the labeled carbon was detected, some of the samples were heated to 320 degrees Fahrenheit—presumably hot enough to kill anything alive inside the soil. If the labeled carbon was being processed by living things, it should stop being released after the sample was heated because the organisms should be dead. On the other hand, if it was being released through a non-living process, then the heating should not have any effect and the labeled carbon should still get released. The samples that were heated stopped giving off labeled carbon—just as Levin predicted would be the case if there were microorganisms living in the Martian soil.

But the other experiments produced results that seemed to suggest there was nothing alive on Mars. For one thing, there were no signs of organic compounds in the soil, the kind of chemicals associated with all life on Earth. And an experiment designed to detect gas given off by organisms as they breathe came back negative. While Levin and a few others remained convinced that Viking had detected evidence consistent with life, the scientific consensus was that there was no definitive evidence of anything alive on Mars.

Both Viking landers also collected data about the surface conditions, soil, and weather on Mars. They recorded temperatures as low as −184 degrees Fahrenheit and as high as 70 degrees Fahrenheit. They found that the winds blow up to seventy-four miles per hour—as intense as a category 1 hurricane, although the low atmospheric pressure would make these winds feel mild to a person standing on the surface. Still, they were strong enough to explain the enormous dust storms that had been previously observed, some of which had completely blocked out the sunlight. Chemical analysis of the soil revealed that it was mostly clay, volcanic in origin, and high in iron, with traces of silicon and aluminum.

Even though the Viking landers were stationary, having them at different spots on the surface was helpful because it gave the scientists a way to estimate how much variation there was from one place to another. The two Viking landers each had a corresponding orbiter that collected data from above the surface. The Viking orbiters analyzed the ice cap at the north pole and determined that it was primarily made not of frozen carbon dioxide, as the Mariner data had suggested, but of water ice. Altogether, the Viking program gave us the most detailed and nuanced view of conditions on Mars to date.

The next step forward, so to speak, was to send robots that could move around the surface. The first of these was part of NASA's Pathfinder mission, which landed on July 4, 1997.[16] Its mobile component, named Sojourner, began exploring the surface two days later. It was little more than a technology demonstration, thanks to its tight budget as an add-on to the Pathfinder mission, and it covered a total distance of only about 330 feet over nearly three months. But the mission showed that it was possible to land a robot on the surface of Mars that could move around and collect data.

Subsequent rovers did much more. In 2004, within a span of three weeks, two rovers, named Spirit and Opportunity, arrived on Mars. Their goal was to explore the surface for ninety sols. Both lasted much, much longer—six *years* in the case of Spirit and more than *fourteen years* in the case of Opportunity. Their long lives meant they could cover a lot of territory. Spirit traversed a total of five miles, while Opportunity traveled an impressive span of twenty-eight miles. By then, the scrappy rover nicknamed "Oppy" had developed quite a loyal fan base on Earth. It had its own social media accounts run by NASA employees, who posted regular updates written from the rover's perspective.

As a dust storm set in on June 10, 2018, Oppy sent its final transmission. Its reliance on solar panels meant that dust storms were a serious threat. The rover could go into hibernation mode to preserve power, but the battery would last only so long. The dust storm went on for two months. Opportunity was never heard from again. Science reporter Jason Margolis, who was covering the story, tweeted, "The last message they received was basically, 'My battery is low and it's getting dark.'"

There was a tremendous outpouring of public mourning over the "death" of the Opportunity rover. Margolis's interpretation of Oppy's last message back to Earth took on a life of its own, becoming a viral meme, even being sold on T-shirts. In his book, *For the Love of Mars*, science historian Matthew Shindell wrote that "rovers and their imaging systems have made Mars seem closer than ever before, and have allowed even greater fidelity in our imaginings of humans on Mars. The public mourning of Oppy speaks to an expectation that robotic exploration of Mars is connected to the human future—to human exploration of Mars, or even human settlement."[17]

If so, what have these robotic expeditions taught us about how we might live on Mars? For one thing, they make it clear that Mars is a place very different from Earth. While the two planets do have some similarities—both are rocky planets, and a day (or sol) is nearly the same length of time on both—in most other ways they are very different. Many of the differences can be attributed to three basic facts that we now know about Mars.[18]

First, its size. Being smaller than Earth means Mars has lower gravity—three-eighths of that on Earth. That means that a person who weighs 185 pounds on Earth would weigh just 70 pounds on Mars. Its lower gravity has contributed to its thin atmosphere, because it's easier for atmospheric gasses to be lost into space. Without much of an atmosphere there's lower air pressure at the surface, which causes liquids to boil at a very low temperature. A person could not survive on the surface of Mars without a pressure suit that also supplies oxygen to breathe. Likewise, although there is water on Mars, it's always frozen at the surface. Most of the water on Mars is buried.[19] Any ice on the surface rapidly turns into gas in a process called sublimation. An example of this was seen when the Phoenix lander arrived on Mars in 2008. It scraped away an area of soft dust, revealing a patch of water ice. While the presence of water ice below the surface had previously been suspected, this was the first time that it had actually been seen. When another photo was taken of the same spot a few days later, a chunk of ice that had broken off when the soil was scraped away was now gone. Rather than melting and turning into a puddle, as might happen on Earth, the ice had been converted directly into water vapor.

Mars's size probably also contributed to the almost complete loss of its magnetosphere. Earth's magnetosphere, the magnetic field that surrounds and protects it from space radiation, is created by the movement of molten metals in its core. Mars is thought to have had a robust magnetosphere some 4 billion years ago when the planet was young. But according to one theory, Mars's smaller size caused the planet to retain less heat than larger planets, like Earth, and as it cooled the metal in Mars's core became sluggish. To make a magnetic field, the metal needs to move from the center toward the edge in a process called convection. The planet's rotation then causes the molten metal to move laterally. As it reaches closer to the surface the metal cools, causing it to fall back toward the center. It's possible that as Mars got colder, its magnetic field was effectively shut off as the metal in its core stopped convecting.[20]

Whatever the cause of its disappearance, without a functional magnetosphere, there is little to stop space radiation from hitting Mars's surface. The constant flow of solar particles, known as the solar wind, eventually stripped the planet of much of its atmosphere. Mars's low gravity could do little to keep the gasses that made up its atmosphere from being driven out into space. The upshot for human visitors to Mars is that the radiation on the surface of the planet is intense because the very thin atmosphere and lack of a magnetic field provide little protection. NASA's Curiosity rover measured an average dose rate striking the surface of Mars in one day as equivalent to about twice the average amount of radiation a person on Earth is exposed to *in a year!*

However, the radiation exposure on the surface of Mars is about half of what a person would be exposed to in space, for example, in a spacecraft on its way to Mars. On the surface of Mars, or any planet, some of the radiation from deep space is blocked by the planet itself. In space, galactic cosmic rays are coming from all directions. Standing on the surface of a planet means that cosmic rays are coming from above but not from below, so the exposure is roughly half as much. Nevertheless, a human could not survive for long on the surface of Mars without some protection from radiation.

The second thing about Mars that makes it fundamentally different from Earth is its position in the solar system. Moving outward from the Sun, Earth is the third planet after Mercury and Venus. Mars is the fourth. Being further from the Sun makes its orbit longer, so as Kepler figured out, a year on Mars is nearly twice as long as a year on Earth. Its distance from the Sun also makes Mars colder, made worse by the lack of much atmosphere to serve as a kind of blanket to retain heat. We now know that the highest its surface temperatures get is a surprisingly mild 70 degrees Fahrenheit. But that's only briefly, during midday, and only near the equator in the summer and in direct sunlight. It's nearly always much, much colder. The average is about –80 degrees Fahrenheit, and a winter day at the poles can reach down to an appalling –225 degrees Fahrenheit.

We have learned a lot about Mars from our long history of observing it from Earth and from the various spacecraft we've sent there. And there have been many surprises. One unexpected finding came from an instrument on the Phoenix lander called the Wet Chemistry Laboratory. When it analyzed a sample of Martian soil, it detected high levels—about half a percent—of a chemical called perchlorate. Subsequent studies have found it to be widespread in the soil all over the planet.[21]

On Earth, perchlorate is much rarer. It's used in fireworks and other explosives, and can be used as rocket fuel. While the discovery of naturally occurring rocket fuel might be good news for the Martian transportation industry, the bad news is that it's extremely toxic to plants, animals, and people. It dissolves in water, and consuming it can affect the function of the thyroid gland, limiting its ability to take up iodide and interfering with the ability to produce the hormones that help control metabolism. Consuming perchlorates can also affect cognitive development in babies and children. Perchlorates would need to be removed from any water collected on Mars in order for it to be safe to drink. The high concentration of perchlorates in Martian soil would be toxic to most crop plants, and any plants that could survive would become toxic for people to eat.[22]

The discovery of widespread perchlorate may help explain the mysterious results from the Viking landers that had seemed at first to indicate the presence of living organisms. Perchlorates can interact with organic compounds to give off gas—including the labeled carbon—in a way that could resemble an organism's metabolism. What's more, radiation can convert perchlorate to hypochlorite, which has the same effect on organic compounds but breaks down at high temperatures. In other words, the results that had convinced Gilbert Levin that there was life on Mars could be explained by chemical processes that have nothing to do with life.

Other surprising discoveries no doubt remain to be made by the robotic ambassadors we've sent to explore the Red Planet. What they find will determine what it will take to make Mars habitable—or whether life does indeed already exist there. That, in fact, is the primary goal of the Mars 2020 mission and its Perseverance Rover.

*

A week after I joined her for the rover planning meeting, I met Kirsten Siebach in her office where she was participating in a virtual science meeting with the other members of the Mars 2020 team. She was sitting at her desk sipping a Diet Coke, facing two computer monitors. Her office was clean and neat, with very little on the walls other than a circular map of Mars. Beneath it, on a table, were two globes of Mars and the LEGO model she had used to show me Perseverance's capabilities. A variety of Mars-themed stickers and posters were pinned to the board behind her computer monitors, including a bumper sticker that said "My Other Vehicle Zaps Rocks on Mars."

A self-proclaimed Martian geologist, Siebach's education and training happened to coincide with the beginning of the Mars rover era. She got started working on Martian geology as an undergraduate student during the time the Phoenix lander was collecting data, and became a collaborating scientist for the Spirit and Opportunity rovers. As a graduate student, she joined the science team for the Curiosity rover, using data she helped collect for her doctoral dissertation at Caltech. Soon after starting her faculty position at Rice, she joined the Mars 2020 science team as a Participating Scientist, becoming one of the most experienced researchers on the Perseverance rover team.

It was a Thursday, and this was the first science meeting that the team had conducted since the planning session I had joined one week earlier. There were seventy people on the call, including us. Ordinarily there would have been a science meeting on Tuesday, but it was canceled because so many of the team members wanted to see the solar eclipse that was visible across much of North America the previous day. Siebach and I had watched it along with many of our colleagues from a ranch in Central Texas. Now it was time to recap what had happened on Mars while we were away over the last week and go over some preliminary scientific findings.

We quickly learned that there had been a problem. It turned out that after all the planning that had been done for Sol 1111, a failure of the Deep Space Network had prevented the instructions from being sent to the rover.

"We spent the whole day [on it]," Siebach told me with a sigh. "People were there until the evening to get the instructions put together. Every detail. Every second. Every data bit. Every watt of power. And sometimes they don't go for one reason or another," she said matter-of-factly.

Without instructions on what to do, the rover just sat where it was on Sol 1111. At least its batteries would get charged as the plutonium in its nuclear reactor continued to decay and very little electricity was used. That meant there was more power available for moving and doing science the next sol. Siebach wasn't sure what had caused the failure, but she took it in stride. Given all of the things that could possibly go wrong with trying to operate a vehicle on another planet, it was reasonable to expect the occasional hiccup.

Unfazed by the delay, the team moved on to describe the progress that had been made and shared a few updates on data. Thanos Klidaras, a graduate student at Purdue University, gave a presentation highlighting some of the recent observations of the view toward the west, in the direction the rover is heading. On Sol 1112, once the instructions had finally reached the rover, the team had gotten their best view yet of Bright Angel, the outcrop they were navigating toward. Siebach pulled up the images on her laptop, and leaned in to get a better view. "That's so cool!" she exclaimed.

The team also got some good images of the valley below the rover, which had been given the name Neretva Vallis. Based on the new images, Klidaras explained that the outcrop at Bright Angel that the team was eager to examine up close appears to extend along the valley, reaching as far as the canyon wall across from the rover's current location at a place called Gardner Canyon.

Based on the new information, the meeting host, Justin Simon, offered the possibility of altering the planned route toward Bright Angel. Rather than continuing along the rocky edge of the valley, where progress had been slow, he suggested that perhaps they could descend to Neretva Vallis where the sandy riverbed would make progress a bit faster and easier. There is also a mound sticking up in the middle of the valley that could be interesting to check out. What's more, they might be able to get a better view of the Gardner Canyon outcrop on the far side of the valley.

The team ultimately opted for the detour. Sixty-five sols later, Perseverance finally arrived at the base of the Bright Angel outcrop. There, the team found something that caught their eye—peculiar shapes that reminded them of leopard spots. The spots were very small, only about a millimeter across, and consisted of irregularly shaped light patches surrounded by a darker colored ring within the typical reddish rock.

"On Earth, these types of features in rocks are often associated with the fossilized record of microbes living in the subsurface," noted astrobiologist and Perseverance science team member David Flannery in a carefully worded NASA statement.

Had Perseverance just detected evidence of life having once existed on Mars?

Scans by the SHERLOC instrument confirmed the presence of organic compounds, the chemical building blocks of life. PIXL analyzed the dark rings and determined that they contained iron and phosphate. These could have been formed by a reaction involving hematite, a form of iron oxide that is abundant on Mars and gives the planet its rusty hue.

"The sample from Bright Angel is exactly the kind of rock we were hoping to find with Perseverance," Kirsten Siebach later told me. "Rover instruments demonstrate that the leopard spots are potential biosignatures, but we need laboratories on Earth to tell us more. I really can't wait until we bring this sample to Earth."

While there was no indication of the presence of any living microbes, the leopard spots at Bright Angel could be the first direct evidence of ancient microbial life. If it turns out that Mars was indeed once home to microbes, the possibility that life could one day return to Mars seems that much more hopeful.

✷

Plans for how people could survive on Mars, whether for short-duration missions or extended stays, have evolved as our understanding of the conditions on the planet has changed.

The realization that Mars is not, in fact, covered by seas, lakes, or even canals—and that liquid water cannot possibly exist for long on its surface—made the idea of living there a lot less enticing. Access to water in the form of ice buried beneath the surface would be a key consideration for any prolonged human presence. There seems to be more ice in the soil near the planet's poles than closer to its equator. On the other hand, temperatures near the poles are so cold that they would threaten not only human activity but also our machinery. Locations near the equator—still cold, but less extreme—seem more promising from a climate perspective but would require locating sufficient quantities of water, perhaps deeper underground.

Early visions of human habitats on Mars often featured large, clear domes built on the surface. In addition to promising amazing views, these attractive-looking domes could serve as greenhouses for crops and other plants—perhaps even recreating entire ecosystems. But the discovery that Mars has both a paper-thin atmosphere and no functional magnetosphere—and the confirmation that radiation levels at the surface are nearly as intense as in deep space—forced a search for habitat designs that would include much more substantial radiation shielding. The easiest and perhaps most effective way to provide shielding from radiation is using regolith, or Martian soil. It could be piled on top of a structure (so much for those amazing views from inside the dome) or the habitats could be built underground. Perhaps the easiest solution would be to create habitats inside lava tubes—natural tunnels that form as lava flows, created as the outer edges cool and solidify while the hot, molten interior flows until it runs out of material.

I have to admit, the idea of traveling all the way to Mars only to live underground strikes me as thoroughly depressing. For our entire evolutionary history, humans have not only lived on Earth, but have lived on its surface. That said, humans have used caves as homes and shelters since prehistoric times, and people even in more recent times have sometimes constructed underground homes. The most extensive example is a network of underground tunnels and chambers in Turkey that was used for at least the last two thousand years. The site, known as Derinkuyu, was discovered by accident in 1963 when a local man was renovating his home.[23] A small hole he found in the wall turned out to be the opening to a tunnel—one of more than 600 entrances to what was once a vast underground community that housed as many as 20,000 people. It consisted of an astonishing eighteen levels, including rooms where livestock and food were kept. Its primary function was so that its inhabitants could hide and be protected from invaders. As such, it had to have its own water supply and to allow air to flow in and out. The inhabitants collected human waste in clay jars they could seal up when they got full, and there were designated areas within the subterranean network for burying their dead.

The description of Derinkuyu reminds me of another type of underground society—those made by ants. I spent years excavating the enormous nests of leafcutter ants for my doctoral thesis. Much like the ancient underground city in Turkey, leafcutter ants construct their nests using a system of tunnels and chambers. A single leafcutter ant nest can occupy some 2,300 square feet in area—the size of a house—and can be home to more than 5 million ants. The ants cut pieces of leaves and carry them into their subterranean chambers where they cultivate fungi that can break down the leaves. The fungi are then harvested by the ants as their primary food source. Tunnels at the edge of the nest bring oxygen-rich air into the nest as the wind blows across it, and large openings at top of the nest release carbon dioxide as the ants and the fungi they grow all respire.

The discovery of the ancient Turkish city seems to suggest that people are able to live underground much like ants. Perhaps Martian habitats can be constructed using architectural designs modeled after ant nests. But there is an important difference to consider. Most ants don't spend their entire lives underground. Worker ants leave the nest to find food, build and repair the nest, and defend it from enemies. Likewise, people who once used caves and other underground homes weren't restricted to them—at least not permanently. Even if they lived underground, which wasn't always the case, they went outside to find food, interact with people from other groups, and—I can't help but imagine—just to get some fresh air and sunshine.

On Mars, going outside would be a lot more complicated. You would need a pressure suit with a life support system to manage your air supply and temperature. And you would expose yourself to high levels of radiation. Surface time would have to be limited.

Another possibility that has been suggested, both by science fiction authors and by scientists, is that we might be able to change the conditions on Mars enough to make it possible to live more comfortably on its surface. In his 1942 short story *Collision Orbit*, science fiction author Jack Williams coined the term "terraforming" to describe the process of transforming a planet to make it habitable.[24] But the idea didn't reach the mainstream scientific community until 1961, when astronomer Carl Sagan suggested, in an article in the journal *Science*, that Venus could potentially be transformed through a process of planetary engineering to prepare it for what he called "comfortable human habitation." This could be accomplished, Sagan proposed, using algae that can survive in extreme conditions.[25]

At the time, Venus was known to have a thick atmosphere made largely of carbon dioxide and high surface temperatures. Just how high was not known, but the Mariner 2 mission just a year later would reveal that it was practically an inferno—its surface temperatures average 471 degrees Celsius. Its atmosphere was shown to be incredibly thick, with clouds of sulfuric acid extending up to fifty miles above its surface. The carbon dioxide acts as a greenhouse gas, trapping heat from the Sun near the planet's surface and causing the planet to bake.

Mars has the opposite problem. With such a thin atmosphere, very little heat is trapped. With Venus looking like a literal hellscape, it didn't take long to apply the idea of terraforming to Mars. After all, we've inadvertently caused our own planet to warm by increasing the amount of carbon dioxide and other greenhouse gasses in the atmosphere. Couldn't we do the same to Mars—deliberately—and make it warmer and more hospitable?

The idea was tempting. After all, it appears that Mars did once have a more Earth-like environment, with mild temperatures and flowing water. Its atmosphere already contains a lot of carbon dioxide, with more thought to be frozen at the poles in the winter in the form of dry ice. We now know that most of the polar ice caps are actually water ice. But based on the data from Mariner 4, which seemed to suggest they were almost entirely made of carbon dioxide, Carl Sagan proposed that the carbon dioxide in the polar ice caps could be released into the atmosphere by darkening them with dust or other dark materials. That would cause them to absorb more heat from the Sun and undergo sublimation, converting the carbon dioxide into gas the same way that dry ice makes white clouds of gas used in fog machines in theaters and spooky Halloween displays. As the atmosphere grew thicker, the greenhouse effect would make it warmer, further melting the ice, which would in turn further thicken its atmosphere in a "runaway greenhouse" effect.

Sagan's idea attracted the interest of other scientists even once the polar caps were known to be mostly water ice with a seasonal coating of frozen carbon dioxide. By 1998 a review of the published research on terraforming Mars, while acknowledging uncertainties, concluded that "the concept can no longer be described as fantasy."[26] In his 1996 book, *The Case for Mars*, aerospace engineer and outspoken advocate for Mars settlement Robert Zubrin explored other options for how to terraform Mars. One possibility he suggested was large mirrors orbiting the planet that could focus sunlight onto the polar ice caps. Zubrin estimated that a mirror 250 kilometers wide would reflect enough sunlight onto Mars's south pole to evaporate the frozen carbon dioxide and water.

Another approach Zubrin offered is to manufacture substances on Mars that are even more effective greenhouse gasses than carbon dioxide and water, like chlorofluorocarbons.[27] Better known as CFCs, these chemicals were once widely used in refrigerants and aerosol spray cans until their harmful effect on the ozone layer was recognized. Zubrin pointed out that not all CFCs destroy ozone, and proposed one in particular—perfluoromethane—as a suitable candidate for terraforming Mars. The raw materials needed for its production are already present there. But, as he acknowledged, it would require huge amounts of energy to produce enough perfluoromethane—comparable to the total energy use of an entire city.

Zubrin listed other—much more destructive—ideas for how to terraform Mars. One involves detonating nuclear bombs on its surface. Nuking the polar ice cap would instantly produce enough heat to vaporize much of the frozen water and carbon dioxide. Additional carbon dioxide and water is contained within the carbonate rocks below the surface that could be released by detonating hydrogen bombs buried all over the planet. But it would take a very large number of H-bombs—millions, by Zubrin's estimates—and then you would be left with a great deal of radioactive fallout that would be spread across the planet by its wind and dust storms. Not exactly what you want to do to a place shortly before you move there.

Another of Zubrin's proposals was to capture an asteroid and redirect it toward Mars, creating an intentional impact. The advantages to this seemingly wild idea are twofold. First, it would instantly heat the planet enough to melt its ice caps and release a large amount of water and carbon dioxide into its atmosphere, jump-starting the desired runaway greenhouse effect. Second, an asteroid could be selected that has some of the elements that are relatively rare on the surface of Mars, like water. After slamming into Mars, the components of the asteroid would be vaporized and incorporated into the Martian atmosphere, thickening it while also improving its greenhouse functionality.

*

To get a better handle on how feasible terraforming might be, I called planetary scientist Jim Kasting at Penn State University, a member of the National Academy of Sciences and an expert on planetary habitability. "I'm a bit of a pessimist—quite a pessimist—on terraforming the whole planet," he told me. "I think it's infeasible . . . I actually don't understand why people are so fixated on terraforming Mars. I think it's because they don't understand how difficult it is," he said.

Kasting walked me through the challenges of increasing Mars's atmospheric pressure and temperature, citing the proposals for nuclear weapons and manufacturing potent greenhouse gasses. He didn't think these were realistic. But then he brought up another challenge I had not yet heard.

"There's another problem that we haven't talked about that may be the hardest one of all, and that's nitrogen. Mars has a teeny amount of nitrogen in its atmosphere," he said.

Here on Earth, the air we breathe contains 78 percent nitrogen—it's the single greatest component of our atmosphere. By contrast, Mars's atmosphere has only about 3 percent nitrogen. Kasting explained that nitrogen is important to have in the atmosphere because it balances out the oxygen we need to breathe without being harmful or dangerous.

"You don't want a pure oxygen atmosphere," he observed. "Pure oxygen atmospheres are very flammable. We learned about that in the Apollo 1 disaster. There was a spark in the capsule and then the whole thing quickly caught fire and burned them up before they could get out. After that, we learned that you have to mix oxygen with another gas, like nitrogen . . . you need about twice as much nitrogen as you have oxygen."[28]

While other gasses, like carbon dioxide or argon, could serve the same function as nitrogen, each of those has its problems. Our bodies cannot tolerate high levels of carbon dioxide. And argon is extremely rare—making up only 1.6 percent of Mars's atmosphere, so it's unlikely there would be enough to buffer the oxygen in a terraformed Mars.

Kasting brought up the idea of bombarding Mars with an asteroid, perhaps one that contains a lot of nitrogen. Most likely it would require many asteroid collisions to supply enough material. In theory, that could work, he thought. "But you'd have trouble keeping it because it gets swept away," he noted. "Mars doesn't have an intrinsic magnetic field today. Its ionosphere interacts directly with the solar wind. That's why it doesn't have any nitrogen—because it lost it by these non-thermal escape processes which would continue to act even as you were trying to terraform it. You're fighting Mother Nature the whole time," he said.

In other words, any nitrogen you put into Mars' atmosphere would not stay there; the nitrogen would be lost to space as the atmosphere gets stripped away by the stream of solar particles that constantly bombards the planet. This struck me as very bad news for the prospect of terraforming.

"That's not unique to nitrogen, right?" I asked. "Wouldn't that be true of any other gas that was added to Mars's atmosphere, including oxygen?"

"Yes. Everything gets stripped," Kasting said.

So even if we could make Mars's atmosphere thicker, increasing its pressure and heating it to the point where it was a place where we could live, the change would be temporary. Eventually, the solar wind would cause Mars's atmosphere to gradually thin, returning it closer to its current uninhabitable state. Terraforming would be an uphill battle, one that would require continual maintenance.

One way that maintenance could be done, Kasting offered, would be to regularly strike Mars with asteroids to replenish the elements in the atmosphere, like nitrogen and water, that were lost to space. But then again, once people are living on Mars, the idea of intentionally blasting it with a massive asteroid would probably be counterproductive.

"The people living there probably wouldn't want to have a several-kilometer-diameter asteroid aimed at them," he mused. I thought back to my canoe trip on the Brazos River. He was right.

It's a sobering irony. One of the primary motivations for settling Mars is to help protect us from the possibility of extinction caused by a disaster on Earth, like an asteroid impact. And yet asteroid impacts might be needed to keep Mars habitable—but they would likely be catastrophic to anyone living on it.

Kasting thinks the science fiction visions of a fully terraformed Mars—in which people can walk around on its surface without pressure suits, and breathe the air the way we do on Earth—is not really feasible. It certainly couldn't be done with our existing technology. Still, Kasting acknowledged that the development of future technologies might change that.

"Everything that I think is difficult might not look so difficult 200 years from now," be offered.

But Kasting pointed out that there are different degrees of terraforming. He admits that a more modest version of terraforming, in which Mars has a thicker atmosphere made mostly of carbon dioxide, might eventually be feasible. With enough carbon dioxide it might be possible to warm the planet to a point where water could flow on its surface. People wouldn't be able to breathe without an oxygen tank and a mask, but they could at least walk around on the surface without a pressure suit. And some plants could be grown on the surface without requiring a pressurized greenhouse—assuming the toxic perchlorates could be removed. The planet would look much more Earthlike, with lakes and rivers surrounded by green hills and valleys.

Another benefit of the denser atmosphere would be some additional radiation shielding, contributing to the habitability of the surface. Without a magnetosphere, there would still be higher radiation exposures than we experience on Earth. And people would still need to live inside habitats with breathable air and some radiation shielding, but the outside environment would not be nearly as hostile.

This vision for the future of Mars seems much more inviting and pleasant than living in sealed containers below ground, as would be necessary without any terraforming. It's a place that would be more likely to appeal to a wide range of people, not only those already eager to live out their sols on the wild galactic frontier.

But what do we actually know about how living in the conditions beyond Earth affects the human body?

2

INTO THE NEW FRONTIER

The shockwave traveled across the nation.

It was October 4, 1957, and the Soviet Union had just stunned the world by launching the world's first artificial satellite, called Sputnik. The Cold War was heating up, and both the United States and the Soviet Union began to see space as the new high ground—the most strategically important position to hold in any conflict. The race was now on to develop the technology to dominate this new frontier. Sputnik consisted of an aluminum sphere nearly two feet across with four long, thin antennae that looked like whiskers. It contained devices that recorded the temperature and density of the upper atmosphere. The antennae broadcast a radio signal including an audible "beep, beep" sound that could be detected anywhere on Earth using a common shortwave radio. What's more, people could look up into the night sky and actually see it pass by.

The idea of a Russian device flying directly overhead was unnerving to many, and some in the US government saw it as an outright threat. Lyndon B. Johnson, who in 1957 was Senate majority leader, argued that if the United States didn't act quickly to get into space, the Soviet Union would take over as the dominant power of the newly emerging space age.[1] The Soviet's ability to send a device like Sputnik into space suggested it could also transport weapons. The USSR had conducted its first test of a hydrogen bomb in 1953. The USSR could now conceivably put a thermonuclear warhead on a satellite and deploy it anywhere in the world. It seemed inevitable that the arena of warfare was shifting to space—and the next logical step was to send people there.

But the reality was that no one knew what would actually happen to a person in space. There were basic questions about whether a human could even survive in a weightless environment. It wasn't clear whether the heart would still work and whether blood would properly flow to the brain and other vital organs. Would the eyeballs pop out of their sockets? Would it be possible to swallow? In addition to concerns about gravity, little was known about how much radiation existed above the Earth's atmosphere and how it would affect the body and mind. There were also concerns about claustrophobia for a person confined to a tiny capsule. On the other hand, would looking out into the infinite void cause a person to panic—or drive them mad?

Nobody had any idea.

The only way to know for sure was to send someone—or something—up there to find out. Both the United States and USSR began research programs aimed at getting answers to basic questions about whether space is survivable. After their success with Sputnik, Soviet Premier Nikita Khrushchev was eager for an encore, and he saw the upcoming anniversary of the Russian Revolution on November 7 as an opportunity to make another statement about Soviet preeminence. That gave his engineers less than a month to prepare. And yet he gave them an ambitious goal—to send an even larger satellite into orbit, and this time it would contain a live dog. But given the very tight timeline for preparations, there were some difficult decisions that had to be made. The priority was a successful launch and orbit. They had not yet tested a way for the satellite to return to Earth in a safe manner. And so, as far as the dog was concerned, this was going to be a one-way trip.

A two-year-old mutt from the streets of Moscow was selected to make history as the first living thing in orbit.[2] Just before the launch, the Russian people learned of the plans through a live broadcast by Radio Moscow from the Institute of Aviation Medicine. The dog that would attempt to be the first living thing to orbit the planet was introduced. As if on cue, she barked into the microphone. From that point on, the dog that had been known to her trainers as Kudryavka, or Curly, began to be known by another name—Laika, which translates as "barker."

The R-7 rocket carrying Laika launched from the Baikonur Cosmodrome in Kazakhstan at 7:30 a.m. Data from Laika's vital signs that were radioed back to the ground showed that her heart and breathing rates soared, as did the rocket. Once it reached the designated altitude, the booster separated from the nose cone containing Laika's capsule. Laika was now traveling at 17,500 mph, faster than any living thing in history. But even more important than her speed was the fact that she was in orbit. As her capsule circled the Earth, she was weightless, floating around in her chamber as much as her restraining device and the confined space would allow.

On each orbit, as the capsule would pass over the Soviet Union, it would send a radio signal to the ground with information about Laika's status. The data indicated that, after the initial high g's of the launch and then the transition to weightlessness, her vital signs began to return to normal as she adjusted to the strange new sensation. But by the third orbit, there were signs of trouble. The temperature inside the capsule was rising. It was now 104 degrees Fahrenheit—way too hot. Laika was barking. By the time the capsule came around for a fourth time Laika had succumbed to the heat.

She had died much faster than expected, but that information wasn't shared until later. The initial news reports focused on the fact that the Russians had successfully launched yet another satellite, Sputnik 2, and that this time it had a living passenger. Before the name of the dog was revealed, the American media had already dubbed it "Muttnik." There was widespread speculation about whether the dog would survive. Eventually, the news came out that Laika had died while in orbit, leading to no shortage of public condemnation from around the world.[3]

While there was never any intention for Laika to return safely, she was expected to last for closer to seven days. Still, the fact that she survived for at least five hours, and that it was apparently heat that had killed her as opposed to something specific to space, meant that from a scientific perspective the experiment was largely successful.

By surviving for hours in orbit, Laika showed that being in space wasn't outright deadly.

Meanwhile, the Americans were conducting animal experiments, too. While none had yet achieved orbit, several species of primates had been launched in rockets that traveled high enough for them to briefly experience weightlessness. Like Laika, many did not survive. Yet the data they provided allowed researchers to conclude that the basic physiological functions necessary for survival could still operate in space. It was time to send a person.

NASA selected Alan Shepard for the coveted title of first human in space. But even though the primate experiments had shown that survival in space was possible, there were still concerns about whether a person would retain enough of their cognitive functions to perform essential operations. They decided to send a chimpanzee first.

And so, on January 31, 1961 a chimp named Ham was strapped into a Mercury capsule at Cape Canaveral on the Florida coast. The name "Ham" was in fact an acronym: HAM, which stood for Holloman Aero Medical—the research lab in New Mexico where he had been trained. His official name was Subject 65, although his handlers knew him as Chang. He was one of forty chimps at Holloman. Others went by names like Duane, Minnie, George, Little Jim, and Elvis.[4]

Ham was three years old at the time of his flight. He had been born in the wild, somewhere in the rainforests of Cameroon in Central Africa. Animal trappers had captured him as an infant, and he was shipped to a private zoo in Miami known as the Rare Bird Farm. The US Air Force purchased him in 1959, at the age of two, for $457. At Holloman Air Force Base, Ham was placed under the supervision of Sergeant Ed Dittmer. Of all his charges, Sergeant Dittmer had a particularly good relationship with Ham. "I know he liked me," Dittmer later recalled.[5] "I'd hold him and he was just like a little kid. He'd put his arm around me and he'd play, you know. He was a well-tempered chimp."

Ham's flight was a success, although it didn't appear that way at first. The rocket veered off course and traveled roughly 33 percent faster than planned, triggering an automatic abort sequence that led to g-forces that made Ham briefly feel seventeen times heavier than normal. Nevertheless, he survived. Sixteen and a half minutes after it had launched, the Mercury capsule splashed down into the Atlantic Ocean. Two hours later the rescue ship, the USS *Donner*, arrived. They opened the chamber to find a safe and healthy—albeit extremely agitated—chimpanzee.[6]

But Ham's flight delayed Alan Shepard's launch, and it cost him the title of first human in space.

On April 12, 1961—just seventy-two days after Ham's flight—a Vostok rocket took off from Baikonur carrying a man named Yuri Gagarin. The Vostok was modified from the R-7 rocket that had launched Sputnik 1 and Sputnik 2. But unlike Laika, Gagarin had a capsule equipped with retrorockets to allow it to reenter Earth's atmosphere—though he would have to eject from it and parachute down to the surface. Gagarin's capsule was also substantially larger, though by no means roomy. Radio telemetry technology would again be used to monitor Gagarin's vital signs during the flight, with the added benefit that Gagarin could communicate directly with the ground team.[7]

"Goodbye, until we meet again, dear friends!" Gagarin shouted into the radio as the Vostok rocket roared into the sky.

Speaking to him on the other end of the radio from a bunker at the Cosmodrome was Sergei Korolyev, a mysterious man known to most only by his title—Chief Designer. He was in fact the head rocket engineer of the Soviet space program, whose identity would be publicly revealed only after his death in 1966.

"How are you feeling?" Korolyev asked Gagarin a minute and a half into the flight.

"I'm feeling good," Gagarin responded. The g-forces were climbing, but his centrifuge training had prepared him well. As the nose cone section separated from the rocket, he suddenly caught his first glimpse through the portholes. "I can see the Earth!" Gagarin shouted in excitement through the radio. "I can see rivers, folds in the Earth. It's all easy to make out. Visibility is good. Everything is wonderfully visible!"

He was still climbing. After a few more minutes he lost radio contact with the bunker, but kept reporting his status. Eleven minutes into the flight the third-stage engine shut down and separated as planned. He loosened his straps and felt himself begin to float.

"The feeling of weightlessness is no problem, I feel fine," Gagarin reported. Then he raised the visor on his helmet to take a good look through the porthole. "I could see the horizon, the stars, the sky," he later recalled. "The sky was utterly black. . . . There were huge numbers of stars. . . . I could see the very beautiful horizon, I could see the curvature of the Earth. The horizon had a beautiful deep blue color. The very edge of the Earth had a soft blue color, which gradually darkened further, turning to a shade of violet, then turning black."

It was a view no human—no living thing, in fact—had ever seen. After all, neither Ham nor Laika had been given a window. For the very first time, a member of planet Earth was looking back at it from the outside.

That view, and the entire experience of being in space, would be life-changing for Gagarin, as it would be for those who would follow. But the fact that he was able to admire it at all—and to describe it coherently through the radio—well, that said a lot. Being in space was a wild ride, to be sure, but it didn't make him crazy. He still had his wits about him.

There was still one more space mystery to solve. He took a squirt of food from a tube like those used for toothpaste into his mouth, and swallowed.

"Swallowing is possible," he wrote in his notebook, just before his pencil floated away.

Gagarin completed slightly less than one full orbit[8] before his spherical capsule began to descend, ripping through the atmosphere toward a rural area near the Volga River in southern Russia. Due to a faulty valve that caused his capsule's retro rocket to shut off one second too soon, the area Gagarin was approaching was 186 miles short of his intended landing zone. He ejected from the capsule at 23,000 feet and parachuted to the ground, landing gently in a freshly plowed field near the town of Saratov. He was off course, but by coincidence it was an area he knew well. He had learned to fly planes in Saratov as a student at the technical college, and it was also the place where he had practiced parachuting as part of his cosmonaut training.

Gagarin had come full circle, in more ways than one.

✷

The United States had lost the race to send the first person into orbit, but Gagarin's flight galvanized the nation to double down on its space program. John F. Kennedy had campaigned for president on the notion of a "new frontier," in which science and technology would lead the way toward an exciting future. And while it wasn't necessarily of personal interest to him—he once told the head of NASA "I'm not that interested in space"—Kennedy recognized the strategic value of America's progress in space exploration. And he understood that there was a lot at stake, in matters of both national security and public perception, in the space race against the Soviet Union.[9]

Sputnik had been a wake-up call, a very public demonstration that the Soviets were ahead in rocket technology. But that had happened during Eisenhower's presidency. It had prompted the creation of NASA and the start of the Mercury astronaut program. But now, on Kennedy's watch, the United States had just been delivered another major blow.

After Gagarin's flight, Kennedy saw clearly the importance of the space race in the Cold War—and to his presidency. Just two days after Gagarin's flight, Kennedy held a meeting with his top space advisors. He was desperate for a way to surpass the Soviets in space. "What can we do? Can we go to the Moon before them?" he asked. "If somebody can just tell me how to catch up. Let's find somebody—anybody. I don't care if it's that janitor over there," he insisted. His vice president, Lyndon B. Johnson, drafted a memo describing where the United States stood in the development of space exploration technology. But the person who would lay out the framework for the type of bold plan that Kennedy wanted was the rocket designer Wernher von Braun, a former Nazi who after surrendering to the Americans near the end of World War II had become the central figure in the US rocket program. Von Braun wasn't at the meeting, but he wrote to the president with his own assessment of what was possible. He thought there was an "excellent chance" that the United States could land people on the surface of the Moon—and that they could do it before the Soviets.

It didn't take long for Kennedy to seize the idea of a Moon landing as the key to American success in the space race. He wanted to show the world that the American model for how to get things done—one centered on freedom and democracy—was superior to the authoritarian, secretive model being used by the communists in the Soviet Union. On May 5, 1961, Alan Shepard finally got his chance, becoming the first American in space aboard the Mercury-Redstone rocket dubbed Freedom 7. He reached an altitude of 116 miles on a ballistic trajectory, meaning that he was shot up into space like a cannonball that went up and then came back down.[10]

Shepard's mission looked a lot like Ham the chimpanzee's flight just ninety-five days earlier. Which, of course, was precisely the point of Ham's flight. The purpose of Shepard's mission was supposed to be to find out what would happen to a human in space, including how being in space would affect a person's cognitive abilities. But, as Shepard was well aware, Gagarin's flight had not only stolen the glory; it had proven that a person could function in space both physically and mentally.

Shepard's achievement was more symbolic than scientific. But it got the nation excited. Some 45 million people watched the dramatic launch on TV, and Shepard became an instant American hero. Kennedy wanted the world to know that this was just the beginning of US efforts in space exploration. Three weeks after Shepard's successful flight, Kennedy was at the US Capitol to speak to Congress.

"I believe the nation should commit itself to achieving the goal, before the decade is out, of landing a man on the Moon and returning him safely to Earth," he declared. It was the first public announcement of his space ambitions, but not, perhaps, the most memorable. That came a year and a half later when Kennedy traveled to Houston, the city that had been selected as the new home for NASA's Manned Spacecraft Center.[11] Speaking at the football stadium at Rice University, he delivered a powerful speech.

"We meet at a college noted for knowledge, in a city noted for progress, in a state noted for strength, and we stand in need of all three," he began. He then went on to frame the nation's space ambitions by placing them within the context of the rapid pace of humanity's scientific and technological developments. The lesson, he explained, is that technological progress is inevitable. If the United States doesn't take the lead into space, someone else will. And everyone in the audience understood exactly who that someone else would be. Anticipating detractors, he delivered the punchline.

"But why, some say, the Moon? Why choose this as our goal? And they may well ask why climb the highest mountain? Why, thirty-five years ago, fly the Atlantic? Why does Rice play Texas?[12] We choose to go to the Moon! We choose to go to the Moon in this decade and do the other things, not because they are easy, but because they are hard," he said.

That last line would become the mantra for Kennedy's "moonshot." He concluded by returning to his metaphor of space as the new frontier, an expansive ocean that awaited exploration much as our ancestors had once explored the unknown lands and seas on our home planet.

"As we set sail we ask God's blessing on the most hazardous and dangerous and greatest adventure on which man has ever embarked."

*

Just a short walk from the stadium where Kennedy delivered his speech, along an avenue shaded by the sprawling limbs of live oak trees, is Rice's Fondren Library where a replica of the lectern that Kennedy used in his speech at Rice Stadium is on display, along with photos of the event. One evening in the spring of 2023, historian Alison Bashford was at the library to give a presentation about a book she had written. It was a multigenerational biography of the Huxley family, which included Aldous Huxley, author of the dystopian science fiction novel *Brave New World,* and his grandfather, Thomas Henry Huxley, a prominent biologist in Victorian England known as a staunch defender of Charles Darwin's theory of evolution. His other grandson, Julian Huxley, was also an evolutionary biologist and was the first professor of biology at Rice University when it opened in 1912.

Prior to Bashford's talk, there was a small reception where a few of the items from the library's Huxley collection were on display. Among them was an original copy of the fifth edition of Charles Darwin's book *On the Origin of Species,* first owned by Thomas Henry Huxley. Delighted at the opportunity to examine such a treasure, I carefully flipped through the yellowing pages. A handwritten note in elegant script on the title page read, "From the Author." Julian Huxley had inherited this volume along with many of his grandfather's books, and Julian had added many of his own notes in the margins throughout the text. I felt my pulse quicken.

As I giddily examined the book, I began chatting with a woman who seemed just as excited about it as I was. She introduced herself as Ann Goldstein. She had white hair and was elegantly dressed, with a long-sleeved pink blouse and a floral-patterned cardigan draped over her shoulders. She explained that she is an Emeritus Professor across the street at Baylor College of Medicine and that she was a Rice alumna, having earned her PhD in cell biology in 1969. I asked about her research area, and she told me that she had spent much of her career investigating the cardiovascular system—including how it is affected by microgravity.

I was thrilled at the serendipity of meeting someone who was interested in the history of evolutionary biology and who had studied how space affects the human body. We immediately made plans to meet again to discuss her work in more detail.

We held our first meeting several weeks later in a room near my office. Ann brought a folder packed with handwritten notes and several photographs. She was eager to show me one photo in particular, a small print that showed a topless young man lying on what looked like an inclined hospital bed. Surrounding him was a room filled with midcentury machinery: oversized switchboards, reel-to-reel magnetic tapes, and bulky monitors. A fashionably dressed young woman was standing next to him, apparently doing something with the complex assortment of mechanical devices.

"That was my Audrey Hepburn phase," Ann said with a laugh. To explain the photo, she first had to give me the backstory. "The summer between my freshman and sophomore years I had a job at MD Anderson Hospital," she recalled. "And I met this amazing Harvard guy who was working in the same lab. I was just typing a manuscript. It was just a summer job. But he was doing electron microscopy. Our first date was actually looking at the electron microscope. In the dark," she said, laughing again.

The experience fueled two passions in Ann—one for the microscopist who soon became her husband, and another for pursuing a career in science. She got a job at the Veterans Affairs Hospital learning about cardiac physiology and then another the Texas Institute for Rehabilitation and Research working on respiratory function. And because she had experience working with both the heart and the lungs, she was asked to help with an exciting new project.

"Carlos Vallbona and Jim Carter were asked by NASA to do a bed rest study," she said. The idea was to examine the effects of prolonged exposure to microgravity on the human body. Kennedy had just announced the plans to send people to the Moon, and there was an urgent need to get a better understanding of how spending more than just a few hours in space would affect the astronauts. Laying horizontally for an extended time without ever getting up essentially limits the impact of gravity, and could provide some insight into how the body would respond. The study would focus on the cardiovascular effects, as well as the effects on muscle and bone.

"The plan was like six weeks of bed rest," Goldstein recalled. But the study had to be cut short. "We didn't do the full six weeks. It would have been unethical. We knew after two or three weeks that this is not good," she said. The patients, volunteers from the military like the young man in the photo, were showing signs of surprisingly rapid deterioration. "There were changes in calcium in the bone, and there was muscle atrophy," she explained. There were also changes in cardiac function being picked up in the electrocardiograph readouts, suggesting the heart was weakening.

The results were unexpected.

"We were so stunned that it was so dramatic," Goldstein added. "We thought we would have to go the full six weeks before we'd see anything." When the table was tilted upright after weeks of lying flat, simulating reentry into the Earth's gravitational field after a prolonged time in space, the heart would suddenly have to work harder to pump blood up to the head. A weakened heart might struggle to keep the brain supplied with oxygen, and that could put an astronaut who had just returned to Earth in serious jeopardy.

The results of the bed rest study suggested that astronauts who were in space for more than two weeks would face an increased risk upon return. That put an upper limit on how long the mission to the Moon could last.

*

With the Apollo missions, the United States finally achieved a dramatic victory in the space race against the Soviet Union. President Kennedy had established the ambitious goal of placing American boots on the Moon by the end of the 1960s. On July 20, 1969—less than six months from the deadline—Neil Armstrong realized that objective when he climbed down the ladder from the lunar lander and set his boots firmly on the gray, dusty surface of the Moon. His first words—spoken to the technicians at Mission Control in Houston and broadcast to 650 million viewers around the world—would go down as among the most famous quotes in history: "That's one small step for [a] man, one giant leap for mankind."[13]

Indeed, it was. Not only was the first Moon landing a much-needed geopolitical victory for the United States, but it was also a demonstration of the incredible resilience of the human body. Given the right equipment, people were capable of traveling through space and even spending time on another celestial body.

Yet only a minimal amount of biomedical data was collected during the Apollo missions.[14] That was not always the plan, however. According to a NASA report, "The Apollo Program originally included the conduct of a series of medical studies." That plan changed after the events of January 27, 1967. Three astronauts—Gus Grissom, Ed White, and Roger Chaffee—were strapped into their seats inside the command module as it sat on the launch pad at Cape Canaveral for a systems test when a fire suddenly burst out inside the capsule. The cabin was filled with a deadly combination of highly flammable material and pure oxygen, the perfect recipe for disaster. The three astronauts were trapped inside, unable to open the hatch.[15]

In the aftermath of their tragic deaths, priorities shifted. After the tragic Apollo 1 accident, the NASA study explains, "the decision was made to delete the medical studies and to dedicate all resources to the complex lunar landing program. Consequently, medical studies were primarily conducted with the Apollo crewmen before and after each flight."[16]

That decision would drastically reduce the amount of information that could be gleaned about the effects of deep space on the human body. Some information about how the astronauts were affected by the flights would also come from a basic bioinstrumentation system worn during flight. It consisted of an electrocardiograph probe to monitor heart rate, a device that tracked their breathing rate, and a thermometer for measuring body temperature.[17] The only additional data came from verbal reports provided to flight surgeons by each astronaut about how they felt before and after the flights. These reports, as with early missions, had to be taken with a grain of salt as the crew members were well aware that any signs of distress could be interpreted by the flight surgeons as a weakness that could jeopardize the astronauts' flight-ready status.

Nevertheless, motion sickness was reported for the first time by American astronauts[18] during the Apollo 8 flight, the first to orbit the Moon. While none of the astronauts who had flown on the Mercury or Gemini missions had reported feeling motion sickness, they may have simply not mentioned it for fear of being grounded. But they had also been strapped into their seats for their shorter-duration flights. The Apollo astronauts, on the other hand, were able to fully experience weightlessness during the six-day flight by unbuckling their harnesses and floating around the command module. The same thing happened during Apollo 9, which lasted ten days but remained in low Earth orbit. According to a NASA report, "Apollo 8 and 9 especially were plagued with vestibular problems: five of the six crewmen developed stomach awareness, three of the six nausea, and two of these six proceeded on to frank vomiting."[19]

Aside from feeling motion sickness, the Apollo astronauts showed signs of dehydration and weight loss after returning home. That could simply be a result of the nausea, given that they did not eat or drink as much as normal. Their postflight blood work showed a decrease in the volume of plasma, which could be a sign of dehydration. But they were also anemic—that is, they had low levels of red blood cells—a symptom of spaceflight that had been detected in the Gemini astronauts as well.

*

The most concerning medical observation during the Apollo missions happened during Apollo 15 in 1971. Astronauts James Irwin and David Scott spent nearly three days on the Moon, including more than eighteen hours exploring its surface on foot and in a rover. During that time, the flight surgeon in Mission Control monitoring Irwin's health noticed a signal from Irwin's bioinstrumentation system that something was wrong. Irwin was experiencing bigeminy—a type of arrhythmia in which the heart contracts an extra time before every other normal beat.[20]

If Irwin felt anything unusual, he didn't say anything about it to Mission Control or to his mission companions. He did report feeling exhausted by the time he and Scott finally left the Moon and returned to the command module. But that was not really that surprising. He had been doing very physically demanding work and had not slept in almost twenty hours. What's more, the water system in his spacesuit had malfunctioned and was not properly cooling during his spacewalks, which lasted up to seven hours each. He was sweating, but without the ability to evaporate the sweat he became increasingly dehydrated, without the benefit of being cooled down.

The folks in Mission Control didn't yet know about the suit malfunction or the fact that Irwin had lost 5 percent of his body weight from the exertion. But based on the cardiac data they were receiving, they were concerned enough about his condition that they encouraged him to take a sleeping pill to help him rest. But not concerned enough, apparently, to have him skip the strenuous spacewalk he did the next day. The spacewalk ended up being successful, and Irwin did not have any more cardiac irregularities during the remainder of the mission. He did, however, have a major heart attack a little less than two years later while playing handball. He survived a second heart attack in 1986 while jogging. In 1991, at the age of sixty-one, James Irwin suffered a third heart attack while biking, and this one proved fatal.[21]

NASA's initial investigation into the cause of Irwin's cardiac episode during his time on the Moon was that it was caused by overexertion.[22] The medical team suspected that all that sweating had caused Irwin to lose a lot of electrolytes, a problem that seemed easily preventable—just give future astronauts a sports drink. NASA developed a special formulation, an orange drink mix high in potassium. It wasn't very popular among those who had to drink it. Apollo 16 astronaut John Young made this abundantly clear when, without realizing his microphone was on during a moonwalk, complained in salty language about the orange drink giving him the farts.[23]

Flatulence was not the only thing happening to the Apollo astronaut's bodies that they weren't eager to share with Mission Control. During the postflight debriefing after the Apollo 11 flight to the surface of the Moon, Buzz Aldrin described a mysterious phenomenon he had experienced on the second night of the flight but hadn't reported until after the flight had concluded.[24]

"I was trying to go to sleep with all the lights out. I observed what I thought were little flashes inside the cabin, spaced a couple of minutes apart and I didn't think too much about it other than just a note in my mind that they continued to be there. I couldn't explain why my eye would see these flashes," he said.

His companion, Neil Armstrong, hadn't noticed any flashes at first, but once Aldrin told him what he had seen, Armstrong started to look for them.

"I'd seen some light, but I just always attributed this to sunlight, because the window covers leak a little bit of light no matter how tightly secured," Armstrong said during the postflight debriefing. "The only time I observed it was the last night when we really looked for it. I spent probably an hour carefully watching the inside of the spacecraft and I probably made fifty significant observations [of light flashes] in this period."

The flashes were caused by high-energy particles—space radiation—passing through the astronauts' eyeballs.[25] The particles were traveling so fast that they passed through the spacecraft and through his body. If they happened to traverse the retina of his eye, the photoreceptor cells would react as if they are detecting light as happens normally when we see.

Here on the surface of the Earth, we are fortunate to be shielded from much of the radiation that exists in outer space. The radiation comes from two sources. Our Sun emits a mostly steady supply of radiation, including the light that is visible to our eyes and mostly harmless, infrared radiation—better known as heat—as well as more dangerous forms including ultraviolet radiation and X-rays. The gasses in the upper levels of our atmosphere absorb the X-rays. Ozone, which is a form of oxygen gas that consists of three oxygen atoms instead of the usual two, absorbs most of the ultraviolet radiation. The result is that only a relatively small amount of ultraviolet radiation reaches the surface of the Earth, along with visible light and infrared radiation. If you're lying on the beach, you really are soaking up the rays—the infrared radiation is what makes you feel warm, and the ultraviolet radiation is what will cause you to get a sunburn if you stay out too long and don't protect yourself with sunscreen.

*

The other source of space radiation was discovered by an Austrian physicist named Victor Hess through a series of clever—and daring—experiments performed in the early twentieth century. Radiation had been detected by instruments at the surface of the Earth, but there was a debate about where the radiation was coming from. Some believed the source was rocks on the Earth's surface, some of which, like uranium, were known to be radioactive. But detectors on boats out at sea had picked it up as well, suggesting something other than rocks were the source of the radiation. Hess suspected they were coming from somewhere beyond Earth and set out to test his idea by getting as far from the Earth's surface as he could.

He wasn't the only one. A German physicist and Jesuit priest, Theodor Wulf, attempted to test whether the amount of radiation changes with altitude by taking measurements at the top and the bottom of the Eiffel Tower in 1910. He found that there was a slight decrease in radiation at the top of the tower, 200 meters above the ground. But the decrease was much less than predicted if the source of the radiation was the rocks in the ground.[26] This was puzzling. Perhaps the Eiffel Tower's metal frame was interfering with the measurements? Or the atmosphere itself was contributing to the radiation?

Hess decided to go higher. Traveling in a balloon filled with hydrogen, Hess took measurements at 1,100 meters but found the radiation readings were about the same as on the ground.[27] Nevertheless, he kept at it, going higher and higher in subsequent balloon trips. Intriguingly, it seemed like the higher he went, the higher the radiation levels he measured. At 5,300 meters above the ground, he found the radiation levels were several times higher than at the surface. This clearly showed that the source of the radiation was not below, but above.

The radiation was coming from somewhere beyond our planet.

One possible extraterrestrial source for the radiation was the Sun. To test that possibility, Hess took measurements from his balloon at night, when any input from the Sun should be much less than during the day when the Sun is overhead. He found the radiation levels were still the same even at night. He even took measurements during a solar eclipse, when the Moon briefly blocks the Sun's rays. There was still no change in radiation readings, proving that the source of the radiation was somewhere out in space, but it wasn't the Sun.

The radiation that Hess had shown was coming from outer space was given a name in 1925: cosmic rays.[28] We now know that the Sun occasionally gives off cosmic rays during major solar storms called coronal mass ejections. But Hess's work showed that there must also be another, more continuous source. Given that the cosmic rays he detected were hitting the surface of the Earth equally at day and night, they had to be coming from all directions. That suggested they were coming from far, far away—from other solar systems, and even from other galaxies. To distinguish them, the cosmic rays coming from the Sun are now called solar cosmic rays and those coming from more distant sources are called galactic cosmic rays.

Galactic cosmic rays are the remains of stars in other solar systems and even other galaxies that violently exploded, sending particles in all directions. Those particles—essentially atomic nuclei with their outer layers stripped off—are traveling near the speed of light. If one were to hit something—like, say, an astronaut standing on the surface of the Moon—it could cause some serious damage. How much damage? At the start of the Apollo program, no one really knew.

The Apollo astronauts were venturing much further into space than anyone had before. The amount of radiation in space was one of the major unknowns at the dawn of the space age. The work of Wulf, Hess, and others had shown that the higher you travel in the Earth's atmosphere, the more radiation exposure you receive. That seemed like an ominous sign for the prospects of the first astronauts, who would far exceed the altitudes that a balloon could achieve. But by the late 1960s both the Soviets and the Americans had sent people into orbit around the Earth, and none seemed to experience any immediate effects of radiation exposure.

But there were also troubling data from satellites that had traveled deeper into space. On January 31, 1958—just three months after the Soviet's success with Sputnik 2—the United States launched its first satellite, called Explorer 1.[29] It contained a cosmic ray detector designed by a physicist from the University of Iowa named James Van Allen and his graduate student George Ludwig. Ludwig and Van Allen received audio transmissions from the satellite when it passed overhead. They could hear the ticking of the Geiger counter, with a faster ticking rate indicating more radiation. The initial results were puzzling. The satellite reached a maximum altitude of 2,515 kilometers, but whenever it was more than about 900 kilometers high, they didn't hear any ticks at all. Could there really be no radiation such a short distance from Earth?

Ludwig and Van Allen wanted more information. Even before Explorer 1 had completed its mission, a second satellite, Explorer 2, was launched to attempt to collect more data. Unfortunately, the fourth stage of the rocket failed to ignite, and the launch was scrubbed. Another attempt took place three weeks later, on March 26, dubbed Explorer 3. This time it was successful. Along with the Geiger counter, Ludwig had included a tape recorder inside the capsule so the radiation measurements could proceed regardless of whether the data could be transmitted back to Earth.

The readings from Explorer 3 revealed why the Geiger counter on the first satellite had gone silent at higher altitudes. Rather than not detecting *any* radiation, it had actually detected so *much* radiation that it overwhelmed the Geiger counter's sensors. The readings were literally off the charts. Van Allen and his team had detected a zone surrounding the Earth where charged particles are trapped by the Earth's magnetic field. They launched additional satellites later that year, probing deeper into space. Van Allen and his team found a second zone located even further away, between 9,300 and 12,400 miles away, where the radiation levels were even higher.

These became known as the Van Allen radiation belts. Each is shaped like a rubber band stretched from both ends then pinched in the middle. The exact size and shape of the radiation belts changes with the amount of radiation that gets trapped. When the Sun is more active, it emits more charged particles and the belts grow. At first, the discovery of the Van Allen radiation belts seemed to suggest that deep space exploration would be too dangerous because it would require passing through both the inner and the outer belts. But the fact that charged particles were being trapped meant that fewer of them were able to reach Earth's surface. In other words, the Van Allen radiation belts were actually *protecting* us from the harmful radiation in deep space.

That's all well and good for those of us on Earth's surface. Anything orbiting close to Earth, lower than 600 miles, is fairly well protected, too. But going farther—including to the Moon—would require passing through the radiation belts. And, what's more, it meant that any place without a magnetosphere would not enjoy the same protection from all that space radiation. Out there beyond the Van Allen belts, out in the vast expanse of space—as well as on the surface of the Moon and Mars—lies a wild frontier where radiation becomes a constant, ubiquitous, and potentially deadly threat.

*

Fortunately, NASA's calculations showed that the Apollo astronauts would pass through the Van Allen belts quickly enough on their journey to and from the Moon that their radiation exposure would not be too high.[30] Just to be sure, each astronaut wore devices to measure their radiation exposure. A personal dosimeter had a readout with a number that showed the cumulative amount of radiation he was exposed to over the span of the entire mission. The astronauts carried one of these devices, which were about the same size and shape as a packet of cigarettes, in a pocket on their flight suit at all times. In addition, they each had several passive dosimeters that they could wear on different locations on their body to provide a more accurate estimate of their total body radiation exposure.

A total of twenty-four astronauts would travel to the Moon during the Apollo missions. Half of them walked on the Moon's surface. None of them experienced any medical problems that could be directly attributed to radiation exposure. The Apollo era proved that human beings, given the right technology and life support, are capable of living in an extraterrestrial world for at least a few days.

But what about longer stays? Could people tolerate space conditions for weeks or months at a time? How would microgravity and radiation affect the body of someone who was actually *living* in space? If humanity was ever going to spend more time on the Moon or explore deeper into space—to Mars, for example—we would need answers to these questions. Answers that neither the Apollo missions nor the Soviet's Vostok program could provide.

To find out, it would be necessary to build spacecraft where people could spend much longer periods of time. As early as 1962, the Soviet Union secretly developed plans for a base on the Moon. It was to be named Zvezda and would consist of a series of interconnected cylindrical modules that could house up to a dozen cosmonauts at a time. The plans called for the base to be covered with lunar regolith for protection from radiation and impacts by micrometeorites. There, they would practice using the Moon's resources to sustain human life while studying how prolonged stays affect the human body.[31]

But the Americans' success in the race to be first to the lunar surface led the Soviets to shift their ambitions away from the Moon. They began plans for a space station that would orbit the Earth below the inner Van Allen radiation belt, where it would be somewhat protected from the most intense space radiation. It would also be faster, easier, and less expensive to send people and supplies to and from the station.

The first space station, Salyut, launched in 1971. It was empty at first. The first crew arrived on April 23, just four days after the station had been launched, but the three cosmonauts were unable to enter because of a problem with docking their Soyuz spacecraft to the space station. After several attempts, they gave up and had to return to Earth.[32] Nearly a month and a half went by until a new crew finally returned to the station. One member of the three-man team who had been scheduled to fly on the second mission was grounded when a doctor noticed a dark spot on his lung during a preflight X-ray. Recognizing the importance of group dynamics on a team that was about to break the record for the longest amount of time in space, rather than replace a single crew member the entire crew was replaced with a backup team that had been training together.

The replacement crew were able to successfully dock with Salyut on June 7. As part of their mission, known as Soyuz 11, they kept track of how their bodies were responding to the environment in low Earth orbit by taking blood samples, measuring their respiratory abilities, and testing their vision.[33] At first, they kept up a regular exercise schedule by running on a treadmill they attached themselves to with tethers, intended to help prevent muscles and bones from weakening in the weightless environment. But running on the treadmill caused the whole station to shake, including the fragile yet essential solar panels and communications antennae. Eventually, they stopped exercising. They were required to sleep in shifts so that one crew member was always awake. The altered sleep schedule and fatigue from the physical and psychological stress of the long mission began to take a toll, and the crew became irritable.

The tension came to a head when the smell of smoke suddenly filled the station. The three men sprang to action, leading to a dispute over protocol and rank that had to be mediated by the ground team. The cause of the smoke—an electrical cable—was identified and repaired. But the commotion caused the crew to argue even more than before, threatening the future of the mission. A moment of apparent camaraderie happened on June 19, when Viktor Patsayev became the first person to celebrate a birthday in space. His companions gave him the ultimate gift for a homesick space explorer: fresh produce—an onion and a lemon—contraband that the others had smuggled aboard. But the tension among the Soyuz 11 crew continued up until they began their return to Earth on June 29 after a record-setting twenty-four days of living in space.[34]

But their return was far from triumphant. As their Soyuz capsule began its descent into Earth's atmosphere, a sudden valve failure caused a leak in the capsule. It was rapidly losing air. Georgi Dobrovolski, Vladislav Volkov, and Patsayev were not wearing pressure suits—there wasn't enough room inside the capsule for them to each wear their bulky suits and helmets. It proved to be a fatal design flaw. The capsule landed as planned, its parachute easing its descent onto the open steppe in Kazakhstan where it was greeted by rescue helicopters. The lifeless bodies of the three crew members were found still strapped in their seats.

They had succumbed almost instantaneously after the capsule began losing pressure. Biomedical sensors on the cosmonauts showed that their breathing rate suddenly shot up when the leak began, and then, less than a minute later, flatlined. There was nothing they could do.

The first people to live in space became the first to die there, too.

3

SPACE MEDICINE

The loss of the three Soyuz 11 cosmonauts as they returned from the first prolonged stay in space served as a harsh reality check about the hazards of space exploration. But prior to the accident in the Soyuz capsule, the cosmonauts' time on the Salyut space station had largely been successful. The crew had completed more than 140 scientific experiments, many of which focused on how the human body responded to spending weeks at a time in space. After mourning the loss of their comrades, the Soviets got back to work planning the next space station mission. The design flaw that led to their capsule depressurizing could be fixed, and in the future all cosmonauts would wear space suits during reentry in case there was another leak, even though that would for many years reduce the number of people who could fit in the Soyuz capsule from three to two.

Much as the Soviet Union had done, the United States in the early 1970s was shifting its space ambitions away from the Moon. NASA had visions of going deeper into space, including plans for a crewed mission to Mars. But after six successful landings on the Moon, public support for the expensive Moon missions had waned. An emergency that nearly killed the crew of Apollo 13, which would have been the third mission to land on the Moon, highlighted the risk associated with sending people into deep space. What's more, domestic concerns, the war in Vietnam, and a greater awareness of the environmental challenges on Earth made it difficult for Congress to justify the enormous budget NASA had been given for the Apollo program. They would have to find more modest ways to advance their space exploration goals.

The compromise solution was to build America's first space station, called Skylab. Like Salyut, it would fly relatively nearby, in low Earth orbit.[1] And to keep costs down it could be made using the leftover components from the Apollo program. Skylab's main living and working quarters, the Orbital Workshop, was built from parts of the Saturn rockets that had propelled the Apollo astronauts to the Moon. In 1973, the first crew spent twenty-eight days aboard the station, exceeding by four days the record for longest spaceflight set by the late cosmonauts two years earlier.

A major focus of Skylab was biomedical research. For the first time, it was possible to do extensive studies on how the human body responds to the space environment. Many aspects of crew health were under investigation, including studies on the heart, blood chemistry, metabolism, bone density, muscle strength, body mass, and the quality and quantity of sleep.[2] The three astronauts acted as both the research subjects and technicians. They took turns doing things like collecting blood samples from one another, weighing their poop, spinning around in a chair to test for motion sickness, and swabbing themselves for studies of microbial growth. At the conclusion of their mission the data they collected were quickly analyzed by teams on the ground, as another crew prepared for their turn on the station. Based on the encouraging results, the next crew—arriving just four weeks after their predecessors departed—was allowed to extend their mission even longer. A total of three crews spent time on Skylab, each exceeding the record set by the previous crew for the longest amount of time in space. The third and final Skylab mission lasted a total of eighty-four days.

The overall picture was encouraging, suggesting that the human body is able to adjust fairly well to being in a weightless environment for months at a time. There is an initial adjustment period that sometimes includes motion sickness that lasts for a few days. Interestingly, being weightless for an extended time caused changes in the astronauts' height. This seems to have led to something of an emotional rollercoaster for some of the Skylab astronauts. "One crewman," a NASA report noted, "is shorter than his wife and was elated to find in-flight that he was finally taller. Postflight there was an undershoot, and he came home to her on the third day postflight shorter than ever." The cause for these changes in height was not yet understood, but was presumed to have something to do with compression of the soft tissues in the spine caused by gravity, which, once alleviated in space, allows the spine to expand.

Another peculiar observation, noted by astronauts in earlier programs and again on Skylab, was that the astronauts' faces became noticeably puffier in orbit. One of them described the experience, explaining that "the eyes turn red which, in my case, happened after about a day or so. The eye sockets themselves become a little puffy, the face a little rounder and a little redder, veins in the neck and forehead become distended and one's sinuses feel congested." Researchers studying the Skylab astronauts determined that these symptoms come from a shift in bodily fluids caused by weightlessness. When we stand up on Earth, the force of gravity causes much of the blood and other fluids in our bodies to be pushed down toward the lower body. In space, without gravity pushing it down, the fluid is distributed throughout the body. The result is legs that appear thinner—what astronauts call chicken legs—and what feels like an excessive amount of fluid in the head and face.

The body interprets the puffy head feeling as a sign that there is too much blood in circulation. It responds in part by sending a signal to reduce the amount of blood. This happens in two ways. First, the volume of blood is decreased by sending a signal to the kidneys to remove fluid from the blood plasma, which is eliminated as urine. Second, a hormone is sent to the bone marrow with instructions to reduce the production of red blood cells. This seemed to explain why astronauts since the Gemini program had been found to be anemic in their post-flight medical exams.

Importantly, there were no major cardiac episodes during Skylab like the one that James Irwin experienced on the Moon. While a few irregular heartbeats were detected, they were not concerning. However, the Skylab astronauts on longer missions found that they felt out of shape when they returned home. Even though they did regular exercise on a stationary bicycle while in orbit, and did not notice any difference in their fitness during the mission, when they got back they were more easily fatigued when exercising. The researchers concluded that this was largely due to the decrease in blood volume that happened in the first four to six weeks as their bodies adjusted to the fluid shifts caused by weightlessness. This also seemed to explain why the Skylab astronauts, upon returning to Earth, had a hard time standing upright. After a couple of months in one g, they were back to their normal fitness levels.

The researchers also noted that based on analyzing their urine, the Skylab crew were losing minerals like calcium, nitrogen, and phosphorus. The loss of nitrogen and phosphorus was interpreted as a result of weakened muscles. Without having to work against gravity, the astronauts lost a significant amount of muscle mass, particularly in their legs. The calcium loss was a sign of bone density loss, which was confirmed with X-rays. Bones grow stronger in response to stresses imposed by the muscles that attach to them. Because the muscles didn't have to work as hard in space, the bones responded by shedding calcium. The authors of the NASA report on the biomedical results of Skylab noted that similar mineral losses were seen in bed rest studies like the ones Ann Goldstein had performed. While the bone and muscle tissue loss did not cause significant problems for the Skylab crew, the researchers pointed out that based on the 0.3–0.4 percent calcium loss per month seen in the three Skylab missions, astronauts on much longer missions could experience dangerous levels of bone density loss.

The Skylab astronauts apparently all slept quite well on Skylab. They found that using a cot with straps that could be used to secure themselves to avoid floating around helped give them the sensation of sleeping in a bed with normal gravity. One Skylab astronaut, Edward Gibson, described what happened when he experimented with not strapping himself down.

"I tried sleeping by just floating free in the workshop," he said. "It was kind of fun, but I could only catnap that way. I floated pretty much with my arms out, as I would in a relaxed position underwater. I'd mash into a wall rather slowly and five minutes later come up against another one. My mind was always half awake, waiting for the next contact. I could never really get a sound sleep that way."

*

The explosion of data from Skylab marked the beginning of a new field: space medicine. Since the start of the American space program, flight surgeons had been involved in evaluating the health of astronauts before, during, and after their flights. Initially, judgments about what an astronaut's body could tolerate were based on experience with military pilots flying jet airplanes, some of which was relevant to space flight.

New information about the unique aspects of space, like microgravity and radiation, had come quickly in the days of Mercury, Gemini, and Apollo. But the flight surgeons had to work with a limited number of measurements taken from a very small number of individuals. Now, with more data, it was becoming possible to ask increasingly sophisticated questions about how space affects the body and to begin making some generalizations.

NASA hoped that having a space station would lead to a sustained presence in space, allowing for even more research on matters related to space, including space medicine. But Skylab began to show signs of wear and tear as soon as it was launched, and while it lasted through its three planned missions as well as a one-month extension, it had to be abandoned after less than a year in orbit.[3] The plan was to move to a new way to access and work in space, a reusable vehicle called the space shuttle. Designs for a space shuttle had begun in the 1960s, in the midst of the Apollo era, but by the time that Skylab was decommissioned the space shuttle was still years away from being ready.

The first space shuttle finally flew in 1981. Having a partially reusable space vehicle—one that could be launched into space attached to a booster rocket, orbit the Earth, and return to the ground again by landing on a runway like an airplane—made access to space much easier and more reliable than ever before.[4] This meant that a greater number and diversity of people could go to space. The shuttle era would last three decades, bringing a total of 355 astronauts to space. They included the first American women, although only forty-nine of the 355 shuttle astronauts were women. That made it difficult to make meaningful comparisons between how men's and women's bodies change in space.

Nevertheless, the available data suggested some trends. A study published in 2014, soon after the final space shuttle mission, found that women tended to have more trouble than men with becoming lightheaded or dizzy after returning to Earth as their bodies readjusted to gravity.[5] This condition, called orthostatic intolerance, is caused by the blood being drawn down into the lower body, which means there is less blood available for the brain. While about 20 percent of men experienced these symptoms after being in space, every single woman who had been to space reported it. Women also lost more plasma from their blood during space flights than men, which made it more difficult to readjust to Earth's gravity because it reduced the total volume of blood they had in circulation when they returned. The authors also found that women were less susceptible than men to developing kidney stones, an extremely painful condition that is more common in space than on Earth because the stones are formed in the kidneys from the calcium that is being lost from bones. Yet there was no evidence of differences between women and men in terms of bone density loss, despite the fact that, as they age, women tend to lose bone density more rapidly than men.

Another limitation in our understanding of how space affects all human bodies was that all of the available information was about healthy adults in the prime of their lives. At the time of their flights, astronauts had ranged in age from the mid-twenties to the fifties. But in 1998, John Glenn traveled to space aboard the space shuttle *Discovery* at the age of seventy-seven.[6] It was not his first trip to space—that had happened more than thirty-six years earlier when he had become the first American to orbit the Earth and the second American in space after Alan Shepard. Glenn was considered a national treasure, not only for his pioneering space flight but also because he was serving his fourth term as a United States Senator. Of course, Glenn's fame made sending him into space somewhat controversial, especially in light of the tragic disaster in 1986 in which the space shuttle *Challenger* exploded shortly after liftoff, killing all seven people aboard. Glenn's age made the flight even more risky, since no one older than sixty-one had ever been to space and so there was no information about how an aging body would react.

But the fact that nothing was known about the effects of space on an older person's body was part of the motivation for sending Glenn. Scott Parazynski, a medical doctor and astronaut, was assigned to be Glenn's personal physician during the spaceflight. He told me about the experience over lunch at a café in Houston. Parazynski said he felt more than a little bit of pressure not only to collect some useful data from his celebrity patient and research subject, but also to ensure he came home safely. Just to be safe, Parazynski brought a defibrillator and a kit stocked with life-support medications with them on the shuttle. He also practiced doing chest compressions in weightlessness—not an easy task—on a parabolic flight.[7]

Fortunately, none of that was necessary. The mission was a success, lasting nine days, during which Glenn was the subject of ten different space medicine studies. They included a study in which Glenn wore a device on his head covered with electrodes to monitor his brain waves while he slept as well as experiments on balance and bone and muscle loss.

"He was a tough-as-nails, marine fighter pilot kind of guy," Parazynski recalled about Glenn, "but he really didn't like needles at all." As part of this research, Parazynski had to take a lot of blood samples from Glenn. Like any good doctor, Parazynski wanted to put his patient at ease. As a joke he wore a pair of vampire fangs during one of the blood draws. It helped lighten the mood, Parazynski explained. "He called me Dracula or Count Parazynski a lot," he said with a laugh.

Overall, Glenn did very well in space. It took him a little longer to fall asleep than is typical for younger astronauts, but he still managed to sleep for about six and half hours each night, which was not outside the normal range for a space shuttle crew member. In general, Glenn's mission showed that older people are capable of adjusting to space in much the same way as younger folks.

*

While NASA was shuttling people back and forth from space for short stays, the Soviet space program was sending cosmonauts on longer and longer missions on their orbital space stations. There were a total of seven different Salyut stations, each slightly more advanced than its predecessor. Salyut 6 and Salyut 7 were the first space stations to have two different places where a vehicle could dock. Since there always needed to be at least one vehicle docked to the station as a kind of lifeboat in case of emergency, having two docking stations meant that one crew could come and go for a shorter mission while another crew stayed on. Using this system the crews on Salyut 6 kept making new records for the longest spaceflight—first ninety-six days in 1978, then 139 days later that year, followed by consecutive records of 175 days and then 184 days in 1979 and 1980. Salyut 7 pushed the record even farther, with two cosmonauts staying for 211 days in 1982 and a crew of three staying for 236 days in 1984.

After Salyut came Mir, a new generation of space station.[8] Not only were there multiple docking locations for spacecraft, but Mir was also modular, so it could be expanded. The first module was launched in 1986 and had room for two crew members. Five additional modules were added over the next decade, each expanding the station's capacity and capabilities.

In 1988, on the last day of August, a Soyuz capsule docked at Mir carrying three cosmonauts. One of them was Valeri Polyakov, a medical doctor who was sent to collect data on the two cosmonauts who had already been on board the space station for eight months. When those two cosmonauts returned to Earth after completing a full year in space—a new record—Polyakov stayed on. By then another two cosmonauts had come to the station, and Polyakov would continue his research by collecting data on them. The doctor returned to Earth himself in April after having spent 240 days on Mir.

Intrigued by the data he had collected on his fellow cosmonauts as well as by his own experiences in prolonged spaceflight, later that year Polyakov took on the role of director of medical support for Mir at the Institute of Medical and Biological Problems in Moscow, where he had worked since 1971. Polyakov had developed an interest in space medicine from an early age.[9] He was born in April 1942 in Tula, where just four months earlier the local people had helped to defend the city from the Nazi army, thwarting Hitler's approach to Moscow. His father died when Valeri was young, and his mother remarried. Valeri changed his family name from Korshunov to Polyakov when he was adopted at the age of fifteen by his stepfather. Just two years later, in 1959 he enrolled in medical school in Moscow. He was inspired by the exciting events in space exploration over the next several years, including Yuri Gagarin's flight in 1961. When the first physician, Boris Yegorov, flew to space in 1964, Polyakov saw an opportunity for a career focused on space medicine. He earned his PhD the following year.

Polyakov's 240 days on Mir only whet his appetite for studying firsthand the effects of prolonged space flight on the human body. He was convinced that the human body could withstand even longer space missions—including a trip to Mars—and he wanted to prove it. He returned to Mir on January 8, 1994. This time he would smash the record held by his former comrades and research subjects. Polyakov stayed on the Mir space station for an incredible 437 days—a new record for the longest time in space that, as of this writing, still stands.

Polyakov finally returned to Earth on March 22, 1995. He was two and a half inches taller than he was at the start of the mission, which was a slight problem because his seat in the Soyuz capsule had been designed to precisely fit his body based on measurements taken before the flight began. Luckily, he could still squeeze into his space suit. After the capsule touched down in Kazakhstan, Polyakov made a point of showing the medical team and others that he could walk without assistance.

"We can fly to Mars!" Polyakov announced triumphantly as he exited the Soyuz capsule. "I was able to come out of the capsule by myself, to walk around, to undress, to dress, to do pretty much everything," he later explained. "And be conscious of everything. That was pretty much the goal of the flight. I had to show that it is possible to preserve your ability to function after being in space for such a long time."[10]

Polyakov's time on Mir had indeed proven that it was possible to live in space for more than a year and still be able to return to Earth without too much difficulty. The studies he conducted during his record-setting mission also helped to reveal some of the ways that the body adjusts to weightlessness. His increase in height, for example, was due not only to compression of the discs that separate the bones within the spine, but also because of a change in the curvature of the spine. Normally, our spines are curved to help support our weight as we walk upright—an anatomical feature that dates back to our ancestors adapting to life on the African savanna. But without gravity pushing down on his body, Polyakov's spine became straighter. After returning to Earth and experiencing gravity once again, his spine would slowly regain its curvature and the discs again flattened, causing him to return to his previous height of six feet two and a half inches.

Measurements of his bone density before and after his flight showed that Polyakov's bones lost about 15 percent of their mineral content. Some bone density loss was expected based on studies from Salyut, Skylab, and earlier Mir missions, but no one knew whether bone density would continue to decline at the same rate during such a long time in weightlessness. The loss of only 15 percent was likely due to Polyakov's strict regimen of two hours of exercise each day.

Polyakov also participated in studies on his mental state during his extended stay on Mir. He recorded his mood and took a variety of cognitive tests before and after his mission, as well as twenty-nine different times over the 437 days he was in space. During the first three weeks of the mission and the first two weeks after his return, there were noticeable changes in his mood. But no cognitive declines were observed, which was encouraging, and his performance and mood remained remarkably stable during the majority of his long mission. Overall, the results were very positive.

*

Mir would remain in low Earth orbit for fifteen years, an era marked by substantial geopolitical change. The Soviet Union collapsed just a few years after Mir's launch, marking an end to the Cold War. Sergei Krikalev, a Soviet citizen, was aboard Mir at the time of the Soviet Union's collapse. He had arrived on Mir on May 18, 1991, and planned to be there for about five months. But the situation back home was rapidly deteriorating. Soviet leader Mikhail Gorbachev resigned on December 26, marking the end of the USSR. The country that had sent Sergei Krikalev to space no longer existed. He ended up staying for a total of 311 days, returning to Earth—and a new country, the Russian Federation—on March 25, 1992.[11]

A new era of collaboration between the United States and Russia had begun, epitomized by coordinated efforts between the two nations in space science and exploration. The space shuttle began making trips to Mir, which had to be modified to allow the shuttle to dock. NASA astronauts would join the Russian cosmonauts on extended stays aboard Mir, and researchers from both nations shared data from studies conducted during the joint missions. But the nations were building toward even greater collaboration—the development of a shared space station.

Plans to build such a station had been in place since 1984, when President Ronald Reagan announced it during his State of the Union address,[12] echoing Kennedy's address to Congress twenty-two years earlier. He envisioned a space station that would have a permanent human presence, and directed NASA to build it within a decade. Unlike Kennedy's directive to put a man on the Moon by the end of the 1960s, this time the president's mandate would not be achieved by the specified deadline. But Reagan's vision for what a space station could achieve—and the collaboration that would be required to achieve it—would provide a blueprint for the next few decades. His directive gave rise to what became known as the International Space Station.

The station's first two segments, one Russian module and one American module, both launched in 1998. They were united together in low Earth orbit, and the first crew—one American and two Russians—arrived on November 2, 2000. The first crew stayed for 136 days, overlapping with a second crew that joined them before the first crew's departure. As of this writing, the International Space Station has been continuously occupied ever since.[13]

The International Space Station would continue to be expanded over the next decade. A second American component, a lab module, was added in 2001. That same year, a large robotic arm was contributed by the Canadian Space Agency. In 2008, the European and Japanese space agencies each added a laboratory module, making the space station a truly international effort that represents twenty-six countries.[14] The completed station was the length of a football field with an internal volume about the same as a Boeing 747 airplane.

Since launches are expensive, the need to consistently have a crew onboard the International Space Station meant that long-duration stays became a practical necessity. In 2012, Russia announced that they would soon send a cosmonaut for a nearly one-year mission. NASA decided to do the same, selecting astronaut Scott Kelly. It would be Kelly's second long-duration stay on the space station, the first of which lasted 159 days. His new mission would be more than twice as long. Kelly admitted in his memoir, *Endurance*, that he had some initial reservations about participating in such a long mission.[15]

"At first, I wasn't sure I wanted it to be me," he wrote. "I remembered exactly how long 159 days on the space station had felt. I had spent six months at sea on an aircraft carrier and that was long; six months in space is longer. Spending twice as long up there wouldn't feel twice as long. I thought—it could be exponential." But Kelly decided he was up for the challenge, and was excited when he found out that he had been selected for the mission. Yet just one day after getting the news, he was told that NASA had reversed their decision after determining that Kelly was in fact medically disqualified for the mission.

The issue had to do with his eyes. During his previous extended stay on the space station, Kelly had experienced some problems with his vision, thought to be related to his time in space. In fact, during his first spaceflight, a one-week mission on the space shuttle *Discovery* in 1999, Kelly had noticed his vision getting worse. "While on mission I realized things were getting blurry in the middle range, ten or twelve feet—across the flight deck of the space shuttle," he wrote. But the problem went away soon after he returned to Earth. But then, during his first long-duration mission, his vision problems returned.

"After a short period in orbit, my vision got worse, and I wore stronger lenses to correct for the change," he recalled. "When I returned to Earth, within a few months my vision returned to what it had been when I left. But I had other troubling signs: swelling of the optic nerve and what seemed to be permanent choroidal folds."

The choroidal folds were especially concerning. The choroid is the part of the eye surrounding the retina, which has all the specialized cells that detect and respond to light. The choroid is filled with blood vessels that supply the retina with oxygen and nutrients. Changes to its shape—like the folds that Scott Kelly had in his eyes—could damage the retina and lead to loss of vision. Similar effects had been seen in some other astronauts during long-duration flights, while others had not experienced any vision problems. NASA thought it would be safer to send an astronaut who had never experienced any vision problems in space for the year-long mission.

But to Kelly, that logic seemed backward. One of NASA's research goals was to better understand what causes damage to the eyes during long-duration spaceflight. Kelly suggested that he might make a good research subject precisely because of the fact that he was susceptible to vision problems in space. NASA agreed, and Scott Kelly was reinstated as the primary astronaut for the year-long mission.

A second reason why Scott Kelly was a good person for the job was that he has an identical twin brother, Mark Kelly. The idea to compare them—with Mark acting as an experimental control while his genetically identical brother flies in space—was actually Scott's idea. As he was preparing for a press conference where NASA would publicly announce that he had been selected for the mission, Scott asked John Charles, the NASA scientist overseeing the project, whether they would be making comparisons with his twin brother.

At first, John Charles told him no.

"I was opposed to it," Charles told me in his office at Space Center Houston in 2019, where he was serving as Scientist-in-Residence. He thought it sounded like a publicity stunt. But after discussing it with colleagues, the team decided it was actually a good idea. Mark was also a NASA astronaut and had been to space on previous missions. But Mark had only flown for a total of fifty-four days, while Scott would spend 340 days in space on this mission, in addition to the 180 days he had previously spent in space on earlier missions. Because Scott and Mark have nearly identical genomes, it would be interesting to observe how Scott's body is affected by a year in space while making the exact same observations of his twin brother's body back on Earth. Any differences between the two brothers might be caused by Scott's prolonged time in space. It was, in simple terms, a test of nature versus nurture—or the relative roles of genetics versus the environment in the functioning of our bodies. With identical twins, they could control for nature (genetics) while testing for the effects of nurture (spaceflight).

NASA's Human Research Program put out a notice that they would sponsor research studies comparing the twin brothers before, during, and after Scott Kelly's year in space. What emerged was the most extensive and complex biomedical research studies conducted in space at that time.[16] The projects included studies of the heart, brain, bone, and muscle; changes to the microorganisms that live in and on their bodies; changes to their cognitive functions, like their ability to perform complex tasks; as well as genetic studies. Samples of blood, saliva, urine, and feces were collected from both twins beginning before Scott Kelly's flight, at multiple times while he was in space, and after his return. In addition, both twins underwent tests of their physiology and cognition before, during, and after the flight.

*

The first results of what became known as the NASA Twins Study were published in April 2019 in the journal *Science*.[17] Some aspects of what they found were expected, confirming observations from previous biomedical studies of spaceflight while providing a more detailed look into what happens to the body during and after a prolonged stay in low Earth orbit. For example, Scott Kelly lost weight while in space, while his brother gained a little. Likewise, markers of bone density loss were higher in Scott than in Mark during the time that Scott was in space. What's more, the choroidal folds in his eyes—the damage caused by his previous spaceflight, which nearly caused him to lose his spot on the yearlong mission—got worse during his year on the International Space Station.[18]

But the Twins Study had some interesting surprises.

One study examined the length of an important section of each twin's chromosomes, called telomeres. Nearly every cell in the body has twenty-three pairs of chromosomes comprising a complete genome—the set of biochemical instructions encoded in DNA for how to build and maintain the body. The telomeres are the sections of DNA wrapped up around proteins at the very tip of each chromosome that serve as a sort of protective cap. Every time a cell divides, splitting from one cell into two, it makes a complete copy of all its chromosomes so that each of the two resulting cells gets a copy.

In fact, the copy of the genome is only *nearly* complete, because a small section of the telomere is lost with each round of duplication. After about fifty cell divisions, the telomeres are gone and the cell can no longer divide. This process is thought to be related to why our bodies begin to break down as we age—our chromosomes are literally getting to the end of the line. Other factors, like stress and disease, can also cause telomeres to shorten. But surprisingly, while the twins had similar telomere lengths before the mission, and Mark Kelly's telomeres stayed about the same length throughout the study, Scott Kelly's telomeres actually got *longer* during his year in space—exactly the opposite of what researchers expected.

Another interesting result was that some of Scott Kelly's cells had evidence of mutations. The sequence of DNA in a chromosome—the precise order of the bases whose names are abbreviated as A, T, C, and G—is what instructs the cell to make particular products. Any change to the sequence counts as a mutation. One type of mutation involves changing just one base, like adding or subtracting an A, for example. Another type of mutation called an inversion happens if a stretch of bases, like the sequence AATC, gets turned around backward to become CTAA. Sometimes a stretch of DNA from one chromosome gets swapped with a section of DNA from another chromosome, leading to a mutation known as a translocation.

Some mutations are harmless, while others can lead to potentially serious problems, including cancer. There is a chance that a mutation could occur every time that cells divide and copy their genomes. But some things can make mutations more likely, including radiation exposure. The Twins Study provided a unique opportunity to find out whether radiation exposure in low Earth orbit causes mutations by comparing the DNA in Scott Kelly's and Mark Kelly's cells before, during, and after Scott's yearlong spaceflight.

Researchers looked for evidence of translocations and inversions by marking DNA on the three largest chromosomes with a fluorescent dye that highlights any section of a chromosome that is out of place, as happens with inversions and translocations. The frequencies of inversions and translocations were similar between the two twins prior to Scott Kelly's year in space, as expected for identical twins. But Scott's rate of both types of mutations increased during his spaceflight, while, back on Earth, Mark's stayed the same.

In other words, there were indeed mutations in Scott Kelly's DNA that happened during his year in space.

In addition to mutations, there were also changes detected in the *function* of Scott Kelly's DNA during his year in space. A major function of DNA is to make proteins, which is important because proteins perform many of the essential tasks in the cells that make up our bodies. A gene is the section of DNA that makes a particular protein. A gene can be turned on—meaning that it is actively making its designated protein—or it can be turned off. There are several ways that genes can be turned off, including by attaching particular kinds of chemicals called methyl groups. This process, called methylation, can effectively silence a gene by shutting off its ability to make proteins. Once a gene has been methylated, it stays turned off until the methyl group has been removed. That can last for the rest of an individual's lifetime—or longer. If sperm or egg cells have any methylated genes when they come together during fertilization, then the resulting embryo will develop with those genes turned off.

The NASA Twins Study was the first to look for evidence of DNA methylation associated with spaceflight in humans. Researchers examined two types of cells, called B-cells and T-cells, both of which are part of the immune system. They found that the pattern of methylation in Scott Kelly's immune cells changed slightly during his flight compared with samples taken beforehand. This suggested that his body was responding to being in space by changing which genes are active and which are silenced. However, the amount of methylation detected in Scott Kelly's immune cells was not different from what was seen in his brother's cells—an example of how useful it was to have Scott's twin brother as an experimental control.

The silencing of genes through methylation is an example of a change that is considered epigenetic, as opposed to genetic. Epigenetic changes are those that occur whether a gene is turned on or off, while genetic changes are those that affect the sequence of DNA, thereby altering the product that is made when the gene is turned on. Genetic changes can be heritable, meaning that the changes can be passed on to future generations, and there is some evidence that epigenetic changes can be, too. While this hasn't been studied yet for astronauts—Scott Kelly had already had kids before his year in space—it has implications for people who might someday live in multigenerational space settlements. Any heritable changes could help them adapt to the conditions on other worlds—yet might also make it difficult for them to return to Earth.

The return is, in many ways, the hardest part of any journey to space. The extreme forces of reentering the Earth's atmosphere—up to five g—can be a shock to anyone, but they are especially hard on a body that has not experienced any weight from gravity for many months. Even normal Earth gravity can be a major challenge at first. Astronauts and cosmonauts are typically helped from their spacecraft after it lands and often need help taking their first few steps. It takes months for their bodies to fully recover—a good rule of thumb is about one week of recovery for every week in space.

Data collected on Scott Kelly right after his return to Earth after nearly a year in space revealed the shock to the system caused by reentry. The levels of many chemicals in his blood and urine that were being monitored before and during his flight showed huge spikes representing dramatic increases or decreases right at the time of his return to Earth. These included markers of bone density regulation, fluid regulation, and overall stress. His immune system seemed to go into overdrive, as indicated by massive levels of chemicals associated with inflammation.

But many of the changes to Scott Kelly's body returned to normal soon after his return. He regained the 7 percent of his body weight he had lost in space. His gut microbiome went back to its preflight composition after just a few weeks. His telomeres shortened, returning to nearly their preflight length just two days after his return. The changes in DNA methylation were reversed, meaning that many genes that were turned off in space were turned on back on Earth. Likewise, measurements of gene activity indicated that many genes that had been more active in space were less active once he was back home. Even some measurements of his eyes—including the thickness of the retina and choroid, which had increased in space—returned to normal on Earth.

Yet there were some changes to Scott Kelly's body in space that did not return to normal even half a year after his return. Many of the mutations in Scott Kelly's DNA—the inversions and translocations in parts of his chromosomes—remained. Likewise, while 91 percent of the genes that had changed their activity levels in space went back to normal, there were still changes in the activity levels of 811 of Scott Kelly's genes after six months on Earth. Some of these genes are involved in the immune system and repairing damage to DNA. The authors noted in their 2019 report that it would be worth looking into these genes in more detail, as they "may be altered for an extended period of time as a result of long-term spaceflight exposure."[19]

*

John Charles, the NASA scientist who oversaw the Twins Study, gave an overview of the study in a presentation in Houston in October 2019. He emphasized the enormous team of collaborators who had worked together on the study, which lasted three years even though Scott Kelly's flight was just one year. Charles joked that the Twins Study even has its own logo, saying that at NASA, "your project isn't beans if it doesn't have a logo." According to Charles, perhaps the most significant takeaway was that spending one year in space does not appear to be twice as harmful as spending six months in space. The body is incredibly good at adjusting to being in weightlessness, and many of the changes happen in the first few months. "Overall, the human body seems very resilient in response to space flight," he said.

Charles acknowledged that the field of space medicine made enormous strides during his career. But he added that, while our knowledge of how space affects the body has increased substantially, there are still major gaps in our knowledge that limit our ability to predict what would happen to people on a mission to Mars. For one thing, a roundtrip mission to Mars is estimated to last about two and half years—more than twice the length of the longest time anyone has ever spent in space. We simply can't be sure how a person's body would react to spending that long in space until someone does it. But based on what we know from Scott Kelly's year in space, Charles was willing to make a prediction.

"I hypothesize that longer-duration spaceflight will have similar effects as shorter-duration spaceflight," he said. In other words, he thought the same kinds of things will happen to the body on a multiyear mission to deep space that we see happening on a one-year mission in low Earth orbit.

As for Scott Kelly, he had a few conclusions of his own. "Personally, I've learned that nothing feels as amazing as water," he wrote.[20] "The night my plane landed in Houston and I finally got to go home, I did exactly what I'd been saying all along I would do: I walked in the front door, walked out the back door, and jumped into my swimming pool, still in my flight suit. The sensation of being immersed in water for the first time in a year is impossible to describe. I'll never take water for granted again."

But there were also other lessons. "I've learned that a year in space contains a lot of contradictions," he added. "A year away from someone you love both strains the relationship and strengthens it in new ways. I've learned that climbing into a rocket that may kill me is both a confrontation of mortality and an adventure that makes me feel more alive than anything else I've ever experienced."

He concluded with an answer to the question that motivated him to embark on the yearlong mission to begin with. "I also know that if we want to go to Mars, it will be very, very difficult, it will cost a great deal of money, and it may cost human lives," he wrote.

"But I know now that if we decide to do it, we can."

*

While Scott Kelly's year in space provided a wealth of information about how the human body might be affected by traveling to Mars, there is one aspect of such a mission that it couldn't quite replicate: the radiation exposure. Because the International Space Station is close enough to Earth to be protected by its magnetosphere, the effects of space radiation from people on the International Space Station aren't the same as the effects of radiation beyond the Van Allen belts.[21] To find out how deep space radiation will affect people, NASA built a device with a name that struck me like something out of science fiction: a galactic cosmic ray simulator.[22]

I had to see it for myself.

After doing some online training and passing a test, I arrived at Brookhaven National Lab on Long Island and checked in at the visitor's center. I was given a visitor's badge and a thermoluminescent dosimeter to track my radiation exposure. The dosimeter is about the size of a name tag and clips on in much the same way. Peter Guida, the lead biologist, met me at the visitor center, and we walked across the parking lot to his office. Guida was sixty-one years old, with dark shoulder-length hair flecked with a touch of gray.

The first thing I noticed about his office is that it was overflowing with plants. "I like living things," he explained. After the plants, I noticed all the pictures of pug dogs. "I don't have kids," he said, "I rescue pugs. But never more than two at a time." In addition to pugs, Guida also has a thing for European sports cars. He drives his Porsche Carrera fifty miles each day to get home, a choice he made to allow his wife, who has a private psychology practice, to be closer to her patients. I got the sense that he doesn't mind the commute.

On the day of my visit there was just one team doing an experiment, along with some time for calibration and scheduled maintenance. Guida gave me a quick overview of the facility and what I was going to see. Pulling out a map, he explained that the entire campus of Brookhaven

National Lab stretches over 5,300 acres of mostly forested terrain. The NASA Space Radiation Lab is only a small section. It was built there to take advantage of the existing infrastructure—namely, the particle accelerators. On the map, I could clearly see several large circles—the facilities where tiny particles are sent around and around at incredibly high speeds. Physicists use accelerators for all sorts of research, including smashing bits of matter together at high speeds to generate even smaller, subatomic particles. Under the right conditions, the accelerators could also be used to simulate space radiation.

"We're stripping electrons off of atoms," Guida explained. He walked me through the basics of how they create charged particles, much like the kind found in galactic cosmic rays. Out in space, all of the elements found on the periodic table from hydrogen to iron are zipping around at nearly the speed of light. They are thought to be remnants of stars that exploded, sending matter racing across the universe in all directions. As they accelerate, they gain energy. The same thing can be done by taking a chemical element and sending it racing around the accelerator.

"Let's say we want iron," Guida said. "We take a little piece of iron oxide. We shine a laser at it. It will start to send off some iron ions. We then pass those ions to the next facility, which speeds them up and gets them circulating in a ring."

To explain how the iron particles become charged, the sports car aficionado offered an automotive analogy. "The best analogy I can think of is water drops on a car after it rains. When you start to drive, some of the water droplets move off. The faster you go, more water droplets come off. That's what we do to get electrons to come off."

It was time to see it for myself. "Put your radiation badge on," Guida advised me as we headed toward the NASA facility. I did. It was about a five-minute drive from his office. I followed Guida's Porsche in my compact Chevy rental car through the wooded campus. The NASA building itself was underwhelming—just a small, nondescript, one-story building with a blue awning covering the door in the middle. I walked inside and Guida instructed me to put on some disposable blue shoe covers like the kind worn in an operating room at a hospital. He introduced me to Trevor Olsen, who had long, blond hair, round glasses, and was wearing comfy-looking slippers.

"So, this is the dosimetry room, where we manage the beam," Olsen explained.

There were desks with several computer monitors, the largest of which had six different windows open showing various complicated-looking displays. On the back wall, above a desk with another computer, were bookshelves filled with binders, hefty textbooks, and a desktop globe of Mars. The wall on the opposite side of the room was made entirely of clear glass, behind which were floor-to-ceiling instruments with wires and lights.

Olsen moved a used plate off one of the desks. "Somebody brought us pie," Olsen said. "We have a tradition where the biology experimenters bring us goodies as sort of an offering to the 'beam gods,' so to speak."

"Is pie a good offering?" I asked.

"Pie is one of the best offerings," Olsen said.

While most of the facility had a decidedly institutional feel, the dosimetry room seemed almost cozy. Colorful paper globes were hung from the ceiling, one of which was yellow and had rays that made it look like the Sun. Any available empty wall space was filled with printouts of schematic diagrams, family photos, a big poster of Neil de Grasse Tyson, NASA mission patches, and a copy of the periodic table of the elements.

Olsen gave me an overview of how they control the type and intensity of charged particles being accelerated to form the "beam." Depending on what is needed for the experiment being performed, the team might use a single type of atom or a mixture, much like what a person in deep space would be exposed to. The lightest atoms—hydrogen—make up 89 percent of galactic cosmic rays. The next smallest particle is helium, which makes up 10 percent of galactic cosmic radiation, while all the heavier elements combined make up just 1 percent. Anything heavier than iron is extremely rare in space, so iron is the heaviest particle used in NASA's simulations.

But being rare does not make a particle irrelevant. Because they are small and traveling at such high speeds, the charged particles that make up galactic cosmic rays pass directly through the tissues of a living thing, like a microscopic bullet. The larger the atom, the more damage it can do. And so, when it comes to radiation, the more relevant measurement is not the *amount* of radiation someone is exposed to but what's called the "dose equivalent," which takes into account not only the amount but also the size, charge, and energy of the particle as well as its biological effect.[23]

"The analogy I always use is a pound of lead and a pound of feathers," Guida explained. "They both weigh the same. But if I drop a pound of feathers on my foot from shoulder height, it won't do much. If I drop a pound of lead on my foot, it might break several toes. That's analogous to dose equivalent. I get much more biological effect from dropping a pound of lead than from dropping the same weight in feathers."

The team had just finished running the beam for an experiment using rats to study the effects of galactic cosmic radiation on the nervous system. The researcher conducting the experiment, Catherine Davis-Takacs, wasn't there that day, but she later told me about what they were investigating. The idea was to mimic the radiation exposure from a mission to Mars by exposing rats to small amounts of simulated galactic cosmic rays over a long time period. While previous research often involved exposing research subjects to a whole lot of radiation all at once—much like what survivors of the atomic bombs in Hiroshima and Nagasaki had experienced—the gradual approach is thought to more realistically simulate the kind of chronic exposure people would experience in deep space.[24]

"Each day, we take the rats out of their individual cages and put them into plastic holders and then we load them into what we call the 'rat hotel,' which holds fifteen rats. It's essentially five rows with three rats per row," Davis-Takacs explained. The entire rat hotel is carried into the target room and secured to the end of the beam line. Once all the people are safely out, the beam gets turned on for the specified amount of time. "The total dose per day is really low," Davis-Takacs said.

Since the rat exposures were done for the day, Olsen offered to take me inside the target room to where the experimental subjects get exposed to the beam. Olsen made a call to an operator to let him know that we were planning to go inside. We got the approval. Leaving the dosimetry room, we turned to the right, where Olsen pulled an access key from a unit on the wall. Another wall-mounted device scanned the irises in Olsen's eyes—a security measure I genuinely thought was only in the movies—and the door to the back of the facility opened. We walked inside. Based on the online training I had done prior to my visit, I knew that being inside the area we were now entering while the beam is on is considered lethal. I felt my heart rate pick up as we stepped inside.

We were now in a winding tunnel. As we came around a bend, we entered a small space with white walls and no windows that was filled with a lot of equipment. This was the target room. The area to our left looked something like a machine shop. The right half of the room had rounded walls and ceiling made of corrugated metal shaped like a tube. We were, in fact, underground. There was a long track with a blue base and round metal rails on the top that seemed to emerge from the wall at the end of the tube part of the room. There were all sorts of devices on top of the track, corresponding to the different places where containers with experimental subjects, like the rat hotel, can be placed.

Olsen showed me one of the ways that the beam's intensity can be adjusted—through lowering one of several rectangular pieces of high-density polyethylene, a type of plastic, in the path of the beam. "We call it a degrader," Olsen explained. "Because we know what the material is, we can calculate how much energy the beam will lose in this material of this thickness."

The blocks of polyethylene were white, but there was a visible dark shadow in the shape of a square where the beam had repeatedly passed through. In the very center of the block was an even darker shadow, caused by the beam passing through when it is more narrowly focused.

Before we left the target room, Olsen had to perform a sweep in which each area of the room is systematically checked to ensure no one is left inside the room. Even though the room was very small and the chances of accidentally leaving someone inside seemed small, given how risky it would be to be trapped inside during a beam run I appreciated the extra safety measure.

The research that Catherine Davis-Takacs is doing at the NASA Space Radiation Lab is revealing insights into how space radiation might affect the brain and cognitive abilities of people traveling to Mars. Studies of astronauts in low Earth orbit suggest that their reaction times can become slower in space. One way that researchers have tested this is by having astronauts use a program on a tablet, which has a small dot or a symbol that briefly appears on the screen. The astronauts are supposed to tap a button when they see the symbol appear. But sometimes, after spending time in space, it can take them a little longer to react. Or they might not notice it at all. This kind of lapse in attention can be an indication that a person's cognitive abilities have been affected by being in space.

As for what exactly is causing the cognitive effect, there are several possibilities. It could just be that they are tired. Astronauts often get less sleep in space than they do at home, and the quality of their sleep can be affected by microgravity or noise. Sleep deprivation is known to affect people's reaction times. It could also be the stress of being in a confined space with just a handful of other people, surrounded by the vacuum of space. Or it could be that the radiation exposure is somehow impacting the brain.

To find out which of these is causing cognitive effects seen in some astronauts, Davis-Takacs designed a psychomotor vigilance test for rats similar to the one used by humans. "Unfortunately, you can't look at a rat and say, 'Hey rat, go press that key and then come back here and press it again,'" Davis-Takacs said. "You have to reinforce them with food."

She came up with a device that consists of a button that a rat can press with its nose. When the button lights up, the rat is supposed to press the button. If it does so quickly, the rat gets a food reward. Just as with human subjects, rats that are sleep deprived or have some other form of cognitive decline show slower response times. The rats in Davis-Takacs' experiments take the test every day after their radiation exposure. What her research has found is that rats that are regularly exposed to simulated galactic cosmic rays show slower response times than rats exposed to other types of radiation or to no radiation at all. In other words, the kinds of radiation that people would be exposed to in deep space, including on Mars, appears to have cognitive effects.

"We need an astronaut to be able to respond quickly at all times," Davis-Takacs said. "Everything they do is life-threatening. And so they have to be able to respond at a moment's notice to any type of emergency. Our data suggests that they're going to be slower to do that—they can still do it, but they're just going to be slower and maybe not as accurate."

The rat studies seem to suggest that the radiation in deep space may affect people in ways that could be significant to their health and safety. The results of her studies make Davis-Takacs worry about how people would be affected while on the Moon or traveling to Mars.

"There's a greater likelihood, I think, for potential accidents, for them not to anticipate an emergency situation as well as they could before, and not to respond as quickly as they would. They'll be a little bit slower, which, in an environment where every second counts, just makes things that much more dangerous," she said.

I left Brookhaven National Lab with a new appreciation for the potential challenges that radiation will pose to our ability to live on Mars. But there was still one more nagging question in my mind—one that could be a showstopper for any efforts to create a permanent, self-sustaining settlement beyond Earth.

Can we have children there?

4

SPACE BABIES

The annual SXSW conference in Austin, Texas, can be a bit overwhelming.

What started as a fairly modest four-day music festival in 1987, drawing some 700 attendees, has become a ten-day extravaganza of panel presentations featuring celebrities and business leaders, film screenings, technology showcases, and—yes—music. These days hundreds of thousands of people converge on downtown Austin for "South By," as it's called by those in the know. When I was a graduate student at the University of Texas in the early 2000s, I always avoided the festival and its inevitable crowds, which was relatively easy to do since it tended to be held the same week as the university's Spring Break.

But when I received an invitation in 2023 to attend a SXSW panel presentation with the title "Sex in Space: Sex and Reproduction Beyond Earth," I knew I had to go. After all, any plans to create a permanent settlement on Mars or elsewhere in space wouldn't last long if we can't have kids there.

The invitation came from a Dutch entrepreneur named Egbert Edelbroek, who would be one of the panelists. I had first met Edelbroek in Houston in 2019, two years after he founded a company called SpaceBorn United. At the time, he was in the early stages of fundraising for an ambitious goal: to pioneer human reproduction in space. Edelbroek came across as a polished and professional businessman, young and good looking, with a big smile and warm personality. He had a PhD in corporate courage development, which seemed appropriate for the bold project he was now pursuing. He told me that the idea for the company had come about after he had become a sperm donor. He was curious about the various ways that donated sperm can be utilized, and he began to learn about assisted reproductive technologies such as in vitro fertilization, or IVF. As a lifelong space enthusiast, who as a child had recurring dreams about being on another planet, Edelbroek wondered if IVF and the subsequent stages of embryo development were possible beyond Earth.

Now, four years after our initial meeting, I was eager to find out what SpaceBorn United had accomplished, so I made the two-and-a-half-hour drive to Austin. The downtown streets were filled with people wearing large, colorful SXSW festival passes dangling from lanyards around their necks. Edelbroek met me at the registration area in the Austin Convention Center. He introduced me to Alexander Layendecker, one of the other panelists and an advisor for SpaceBorn United. A helicopter pilot in the US Air Force Reserve, Layendecker earned a master's and a doctorate from the Institute for the Advanced Study of Human Sexuality. We walked down the hall where we met the other two panelists.

Simon Dubé was a postdoctoral researcher at Indiana University's Kinsey Institute where he was studying human sexuality with a focus on "human-machine erotic interaction." The fourth panelist was Shawna Pandya, a Canadian emergency room physician with a focus on extreme environments and a real passion for space. Indeed, Pandya seemed to be everywhere in the worlds of space medicine and space science. I had met her briefly at NASA's Human Research Program Investigators Workshop in Galveston a few months earlier, and her name popped up as a speaker at almost every other space-related conference or event I learned about or attended.[1] This was shaping up to be an intriguing presentation.

After a quick lunch we walked a few blocks to the JW Marriott hotel where the panel would be held. I accompanied Edelbroek and the other panelists to the green room to prepare. While Dubé, Layendecker, and Pandya sat casually chatting around a table, I noticed Edelbroek was sitting by himself in the corner of the room practicing his presentation. He seemed nervous. This was a high-profile public event and a big opportunity to share what his company was doing. Soon, a woman with a headset appeared and told us it was time to head to the stage. She escorted us down a back staircase, through a service hallway, and into a room filled with people.

Layendecker appeared first on stage and introduced himself as a sex researcher and astrosexologist. He then introduced his fellow panelists one at a time. Upon taking the stage, Shawna Pandya thanked him and added, "I hope we have an uplifting discussion with a happy ending."

Clearly this was not going to be a dry, academic presentation.

Each of the panelists said a few words about why research on sex and reproduction in space is important.[2] Pandya argued that research on human reproduction in space not only is essential for our long-term survival, but is also a short-term need. Space tourism is on the rise, and with the expected development of space hotels in low Earth orbit, she pointed out that understanding the consequences of a human baby conceived in space has become critically important. "We know what people do in hotels," she quipped.

The conversation shifted toward a discussion of whether anyone had already had sex in space. The official answer from government space agencies like NASA was that no one has—but that hasn't stopped the speculation. The people perhaps most likely to have completed what Pandya called 'the first human docking maneuver in orbit" are Mark Lee and Jan Davis. In 1992, they spent eight days together on the space shuttle *Endeavour* as newlyweds. NASA didn't typically assign couples to the same mission, but Lee and Davis kept their marriage a secret until shortly before the launch, allowing them to be the first couple to spend their honeymoon in space. Afterward they declined to answer any questions about what, if any, private interactions they had while in orbit, and NASA certainly would not entertain any suggestions that their astronauts did anything "unprofessional."[3]

Some speculation has existed about whether the basic physiological processes necessary for intercourse, particularly for men, might be compromised by microgravity. But NASA astronaut Mike Mullane put this concern to rest in his 2006 memoir, *Riding Rockets*.[4] "To my surprise, I did not wake up alone," he wrote. "My closest friend was alert and waiting. I had an erection so intense it was painful. I could have drilled through kryptonite. I would ultimately count fifteen space wake-ups in my three shuttle missions, and on most of these and many times during the sleep periods my wooden puppet friend would be there to greet me."

So much for that potential problem.

What about for women? No reports of female astronauts have yet included any description of how their time in microgravity affected their libido or sexual response. But, as is true anywhere, human reproduction in the cosmos will depend on more than just the act of sexual intercourse. Another concern is how the gametes, or the reproductive cells, might be affected by conditions beyond Earth.

A 2022 paper by Stefan du Plessis and colleagues reviewed all the available scientific literature on the topic of how space conditions affect sperm motility and other factors related to male fertility.[5] They found a total of twenty-one relevant articles that used a variety of techniques to assess the effects of microgravity and/or radiation on sperm—though only one study examined the combined effects of both microgravity and radiation. Only four of the experiments were conducted in space. Four of the studies involved human sperm—none of which were conducted in space—with the remainder involving animal research subjects. The authors concluded that the limited amount of available evidence makes it difficult to conclude whether spaceflight has a positive, negative, or neutral impact on the movement abilities of sperm. However, they did find evidence that sperm production decreases when exposed to microgravity and radiation. What's more, radiation exposure can cause harmful mutations to the DNA in sperm—mutations that have the potential to be passed on to subsequent generations.

Given how much biomedical research has been conducted in space, the scarcity of research on reproduction surprised me. Indeed, a report released in 2023 by the US National Academies of Science, Engineering, and Medicine pointed this out, concluding that our understanding of how space affects reproduction is "vital to long-term space exploration, but largely unexplored to date."[6]

Addressing the other panelists at the SXSW event, Alexander Layendecker asked why so little research has been done on sex and reproduction in space. Shawna Pandya pointed out that there have in fact been reproductive studies on a wide range of animal models, including rodents, jellyfish, quail, sea urchins, and fish. But the data, she noted, were conflicting.

Layendecker agreed, pointing out that NASA has historically discouraged its employees from publicly discussing the topics of sex and human reproduction in space. He cited the example of Yvonne Clearwater, who led NASA's habitability research group in the 1980s. While working on designs for space habitats, she advocated for including sleeping spaces with privacy and soundproofing.

"If we lock people up for 90-day periods, we must plan for the possibility of intimate behavior," Clearwater wrote in a 1985 article for *Psychology Today*.[7] The backlash was severe, Layendecker said. Media headlines implied that taxpayer dollars were being frivolously spent on promoting sex in space, prompting NASA to have to assign someone to reassure Congress that this was not the case. According to Layendecker, Clearwater was subsequently forced by NASA to publicly state that no work was being conducted to promote or encourage sexual activity among astronauts. "So politically, [sex] has always been a very explosive topic in NASA," Layendecker concluded.

"Fortunately, they realize that," added Egbert Edelbroek, noting that NASA officials recognize that the political climate limits the agency's ability to fund and support research on sex and human reproduction. "And they explicitly encourage focused biotech companies like SpaceBorn United to address this challenge," he said. Edelbroek suggested that to advance our understanding of human reproduction beyond Earth, it would be helpful to separate sex from reproduction. That way, he argued, each stage in the process of human reproduction can be carefully studied to determine how they are affected by the conditions in space.

"For the next five to ten years we will focus on the very first stage of reproduction . . . conception and the first five to six days of embryo development," Edelbroek said. He stood up and walked over to a table next to the platform where the panelists were sitting and removed a white tablecloth to reveal what looked like a UFO. It was a disc-shaped object, mostly white with a black ring around its wide base. It tapered slightly toward its flat top, giving it the appearance of a flying saucer.

"This is where the actual magic is going to happen," Edelbroek announced. The object was a full-sized replica of the space capsule that will house a miniature IVF lab. A disc inside the mini lab can be rotated at various speeds to produce artificial gravity to replicate conditions on the Moon or Mars, he explained. Using microfluidics, a chamber containing sperm can be connected to a chamber containing an egg to allow fertilization.

"This capsule has return capabilities. It has a heat shield and a parachute system. It can bring five-day old embryos back to Earth," Edelbroek said.

Under normal conditions, a fertilized egg will begin to divide after about twelve to thirty-six hours. After another fifteen hours or so, each of the resulting cells will divide again, such that now there are four cells. The divisions continue, with the total number of cells in the embryo doubling each time. These early cell divisions occur so rapidly that the cells don't have time to grow before they split. That means that the size of the cells decreases with each round of divisions. After seven rounds the embryo consists of a ball of 128 tiny cells with a hollow space in the center. Subsequent divisions lead to multiple layers of cells, which begin to differentiate into those that will eventually form the fetus and those that will form other structures, like the placenta and umbilical cord.

During these early stages of development, the embryo is normally moving through the Fallopian tube. After about five to six days it reaches the uterus, where it implants and completes its development. By this time the embryo has reached the hollow ball stage, called a blastocyst. During typical IVF, an egg is fertilized with sperm in a culture dish where it is allowed to develop into a blastocyst. The embryo can then be frozen until it is time to introduce it to the uterus. SpaceBorn United plans to allow fertilization and blastocyst formation to occur for about twenty embryos at a time in a three-dimensional growth medium while the device is in space, after which the embryos will be frozen. As Edelbroek explained to the SXSW audience, freezing the embryos will pause their development and protect them during their return to Earth.

"We need to protect them and we do that by cryogenically freezing them in exactly the same way that it's being done on an everyday basis in IVF clinics to safely bring them back to the IVF clinic to study their health," he said. "We have our first prototype finished and it's going to go on board a rocket this year within six months."

The crowd burst into applause and, for the first time since the panel presentation began, Edelbroek smiled.

*

I left Austin even more curious about the possibility of human reproduction in space, particularly the ambitious timeline that Edelbroek had announced. But the plan to test their prototype later that year turned out to be overly optimistic. The company that had been contracted to perform the launch in Iceland was not able to get the necessary launch permits in time. After additional delays, the first test launch of their prototype finally took place on a SpaceX rocket launched from Cape Canaveral in 2025.

SpaceBorn United has been one of the only groups attempting to study reproduction in space. In part, that is because NASA, the European Space Agency, and other governmental organizations have been reluctant to fund such work. The National Academies report pointed this out, and recommended a tenfold increase in funding for biological and physical sciences—including studies on reproduction. I spoke with Robert Ferl, who cochaired the group that produced the report. He told me that research should include studies on reproduction in many different organisms, since the underlying biological principles are largely the same as in humans. "We've got to know what happens over generations, because there are fundamental processes involved when an egg is produced, when sperm is produced, and when the new [embryo]—no matter what organism it is—begins to grow and develop," he said.

Some researchers believe that the work by SpaceBorn United will have positive outcomes. Jeffrey Alberts, a professor at Indiana University who has studied the effects of spaceflight on rodents, told me that he was optimistic. "I've come to the general conclusion that fertilization [in space] will probably work," he said.

But fertilization is just the first part. For Edelbroek's team to study the effects of spaceflight on fertilization and the first few stages of cell division, the embryos have to get back to Earth. Dorit Donoviel, director of the Translational Research Institute for Space Health at Baylor College of Medicine, told me that she was concerned about the reentry part of SpaceBorn United's plan.

"Those blastocysts are going to experience massive g-loads coming back," she said. Donoviel told me she was also worried about the legal aspects of research done by private companies like SpaceBorn United. She coauthored an opinion piece published in *Science* in 2023 that argued for more stringent and consistent guidelines for research in the commercial space industry.[8] Donoviel told me she was particularly concerned with SpaceBorn United's long-term plans to conduct IVF experiments in space using human embryos. She said she considers this unethical and worries that it could turn public opinion against all types of space research. "It extends a negative aura over our entire industry and field, so I'm very much against this work," she said.

I asked Edelbroek what he thought about Donoviel's concerns during a Zoom call several weeks after the SXSW presentation. He told me that SpaceBorn United is taking ethical issues very seriously and that they recently added two biomedical ethics advisors to their team. He also emphasized that they will follow all international legal and ethical standards for using human embryos.

But experiments on reproduction do not necessarily need to involve human samples. Jeffrey Alberts told me that he wants to see several generations of animals like rats be born in space, live their entire lives there, and reproduce. Such experiments would be the definitive test of whether there are any multigenerational effects of life in space—an unanswered question highlighted by the National Academies report. The results of such studies would reveal a lot about whether space settlements could ever become a reality. But to Edelbroek, the fact that multigenerational studies on animals have never been approved is the raison d'être for his company. And while its research might make some people uncomfortable, he sees pushing the boundaries as important.

"Humanity has benefited all the time from expanding her comfort zone," he said. "And if you ask me, it's good to continue to do that into space."

Edelbroek hopes to show that the first steps in the formation of human embryos can happen under the partial gravity conditions found on the Moon or Mars, which would be a major step toward demonstrating that space settlements are feasible. But we still need to know about the later steps in fetal development, like the formation of organs, the skeleton, and the brain.

The first experimental test of all stages of development, from fertilization to birth, was performed aboard space shuttle *Columbia* in 1994.[9] The experiment involved Japanese rice fish, known as medaka, that flew aboard the shuttle in a small aquarium. According to the biologist who designed the experiment, Kenichi Ijiri, previous experiments had shown that fish usually swim in a "looping" pattern in microgravity. However, through experiments on parabolic flights they found that some individual medaka swim normally in reduced gravity. Ijiri and his colleagues allowed these fish to breed with one another and produced a strain of medaka that always swim normally in microgravity. From this strain, they selected four individuals—two males and two females—for the experiment aboard the space shuttle.

On the third day of the mission, Japanese astronaut Chiaki Mukai noticed three tiny eggs in the aquarium. By the next day there were ten. A total of forty-three eggs were laid over the span of the fourteen-day mission. On day 12, one of the eggs hatched and Mukai recorded a video of the tiny little fish—the very first vertebrate animal ever born in space—swimming around the enclosure. Soon there were seven more. All eight baby fish survived the return to Earth, and the researchers noted that they seemed to behave normally. The results seemed encouraging for the prospect of reproduction beyond Earth.

Video taken during the spaceflight captured the adult fish in the act of mating—the first known instance of sex in space. The video also allowed researchers to observe the development of the fish through all stages of embryo development until the baby fish emerged. The fact that medaka fish lay eggs that develop outside the mother's body—and that the eggs are transparent—is a big advantage for studying their development. But it doesn't necessarily give us much insight into how the conditions of space would affect development in humans. To get a better sense of what would happen during a human pregnancy in space, we need to look at research on reproduction in mammals.

*

In 1983, researchers from the Institute of Biomedical Problems in Moscow sent ten pregnant rats into space on a Soviet satellite known as COSMOS-1514. At the time of launch, each of the rats was thirteen days into their pregnancy, which usually lasts a total of twenty-one days. This corresponds to the beginning of the third trimester, when the fetus is developing its bones, muscles, and organs. The researchers described this as the "most stable" stage of the pregnancy, which is why they chose it for the first-ever experiment on the effects of spaceflight on the fetal development of a mammal.[10]

The researchers noted that the pregnant rats ate as much food as pregnant rats housed on Earth under similar conditions as an experimental control group. Yet the rats that flew in space gained much less weight—only five grams on average, compared with sixty-five grams in the control group. There were other differences, too. The space rats had lower hemoglobin levels and their livers and thymus glands shrank, while their adrenal glands grew somewhat larger.

Five of the pregnant rats were sacrificed and dissected after returning to Earth to examine the status of the fetuses developing in their wombs. All the fetuses—now in their eighteenth day of development—had survived. However, some of their placentas had experienced potentially dangerous hemorrhages thought to be caused by the high gravitational forces during reentry.

The remaining five females were kept alive and completed their pregnancies. All five gave birth, but the pups in one of the litters were stillborn. The death of the stillborn pups was attributed to complications during delivery resulting from a single large pup—the first to attempt birth—becoming stuck in the birth canal. This ordeal lasted more than a day and resulted in the other pups, which were apparently healthy, to die from a lack of oxygen. Like all astronauts returning to Earth from space, the mother was exhausted and weak. Her weakened muscles, including perhaps those of the uterus that contract during labor and delivery, may have contributed to her complications.

Even the pregnant rats that gave birth to live pups had longer deliveries than usual. Among the pups that seemed to be born healthy, about one in five died after less than a week. Compared with the pups of pregnant rats in the control group that were housed in similar conditions on Earth, the pups that had been in space were smaller at birth. Based on the dissections of the fetuses that were sacrificed, their bones were less fully formed than those in the control group.

Given that bone density loss is one of the well-known consequences of microgravity for adults, perhaps it makes sense that the bones of a developing fetus would also be affected by weightlessness. This could pose a major problem for space settlements. If human babies were born without fully formed bones, it could increase the risk of infant mortality and might affect subsequent bone development. What happens to a baby born on Mars who has delayed bone development at birth and continues to live in a three-eighths-g environment throughout their childhood? Would their bones ever become fully formed? Perhaps some of their bones would develop normally while others would be weakened?

To find out more about how bone density loss could affect reproduction in space, I visited Jennifer Fogarty, who had replaced John Charles as the Chief Scientist for NASA's Human Research Program. In her office on the seventh floor of Building 1 at Johnson Space Center, I asked her whether bone density loss in microgravity affects all bones equally. She explained that, according to their data, it does not. Certain bones, like the femur, lose a much greater percentage of their density during spaceflight than do other bones.

"What about the pelvis?" I asked, thinking about how it plays such a major role in bearing the additional weight during pregnancy. Fogarty explained that there has been less direct research on changes in bone density on the pelvis. In part, this is because measuring bone density requires X-rays, and subjecting the pelvis to repeated X-rays means exposing the reproductive organs to potentially risky levels of radiation. Much of the data on bone density loss comes from X-rays of the wrist or heel, which are easy to access and serve as a proxy for all the bones of the body.

Still, enough research on variation in bone density loss in astronauts has been done to draw some general conclusions.[11] One of the bones that shows the greatest amount of density loss in microgravity is the femoral head, the ball-shaped knob that forms part of the socket joint in the hip, connecting the upper thigh to the pelvis. Fogarty showed me some X-ray images of astronaut's hip bones before and after spaceflight.

"You get a sense for density by the opaqueness," she explained, while gesturing to the black-and-white image on her computer screen. "You see how there's almost no cortical?" she said, pointing to a spot on the image. "It's all trabecular. Right up to the edge."

Cortical bone is the dense, outer part of bones that gives them much of their strength. In contrast, much of the interior of certain bones, like the vertebrae and the long bones of the arms and legs, is made of trabecular bone. Although trabecular bone plays a role in strengthening bones, it is much less dense because it has large open spaces called pores, causing it to resemble foam. Cortical bone has pores, too, but they are very tiny, giving cortical bone the appearance to the naked eye of being solid.

Having these two kinds of bone was a solution to the problem our early ancestors faced as they evolved from fish that lived in the water to animals that lived on land. The buoyancy of water counteracts the force of gravity, which is why astronauts often train for microgravity in giant swimming pools. It might also explain why the medaka fish didn't experience issues related to bone development in microgravity, while the rats did. Being out of the water exposes the body to the full force of gravity.[12] During their evolution from fish ancestors, the first four-legged land animals had to develop stronger skeletons to support their newfound body weight. But having thicker bones adds weight, which further increases the need for stronger bones, in a vicious cycle. The solution was for the bones to be somewhat hollow, with pores that make them lighter, particularly in trabecular bone.

The pores of both cortical and trabecular bone can expand under certain circumstances, such as when the bones aren't being regularly compressed, which makes them weaker. As we age, we naturally lose trabecular bone and get larger pores in our cortical bone as well, leading to osteoporosis and a greater risk of fractures.[13]

The X-rays of the astronauts' hips showed a distinct decrease in cortical bone, consistent with a weakening due to the lack of gravity pushing down on the body while in space.[14] While some have likened this weakening in spaceflight to accelerated aging, Fogarty told me that she doesn't find that comparison to be accurate. For one thing, she pointed out, for astronauts the bone density loss is reversible. Once they return to Earth, they can regain bone mass. Still, there are similarities between the weakened bones of the elderly and those of astronauts after a long spaceflight.

"Does that make astronauts more likely to have broken bones after they return to Earth?" I asked.

"We've had two crew members post-flight break hips," Fogarty told me. Apparently, in both cases, the former astronauts had become injured due to a fall that was unrelated to their spaceflight. "This was several years post-flight," she added. "We could not tie that back to their spaceflight experience."

Indeed, according to the available data, astronauts are no more likely to break a bone following spaceflight than the average Earthbound person. I found this somewhat surprising given what we know about how their bones are affected by microgravity. Numerous studies of astronauts have measured how quickly they lose bone density. According to a 2009 study published in the journal *Bone,* when measuring the amount of bone mineral per unit area using dual-energy X-ray absorptiometry (DEXA), astronauts lost about 1 percent per month from their vertebrae and 1.6 percent per month from the head of the femur.[15] When measured another way, using a technique called volumetric quantitative computed tomography, the total volume of bone mineral density in the head of the femur decreased by about 1.5 percent per month. The loss of mineral density of trabecular bone was even higher, as much as 2.7 percent per month. This is twice the rate at which trabecular bone is lost in women over the age of fifty, the population most prone to osteoporosis.

But Fogarty reminded me that astronauts tend to be very health-conscious people who have excellent bone health prior to their flights, so any bone density they lose in space is unlikely to place them at high risk for fractures. Perhaps they are also especially cautious after returning to Earth. Quite likely the selection criteria for astronauts combined with their strict exercise regimens speed up their recovery as well. Indeed, the bone mass of astronauts' femoral heads was found to return to what it was before spaceflight after just one year. But interestingly, the increased mass was mostly due to the bone growing in size, not recovering the mineral density. That's especially true of trabecular bone, Fogarty told me. "Once the trabecular connections are broken, they don't reform," she said. In other words, their bones got bigger after returning to Earth but not necessarily stronger.

What does all this mean for people living in space settlements? How would their bones respond to the lower gravitational force on the Moon or Mars, and how would this affect the abilities of women to tolerate pregnancy and childbirth? Fogarty agreed that understanding the ways in which adapting to the conditions in a space settlement affect pregnancy, childbirth, and child development are essential. Yet very little is currently known.

"These questions need to be addressed," she agreed.

Indeed, people living their entire lives in a lower-gravity environment would have significantly weaker bones by the time they reach adulthood. Anyone born and raised on a space settlement would have a substantial risk of fracturing bones if they were suddenly subjected to strong forces—forces like those of childbirth. Bone fractures during childbirth are rare, but they do happen. Women with lower bone density are more susceptible to fractures during pregnancy and while giving birth. Women often lose bone density during pregnancy, as calcium from the mother's bones is absorbed by her body and transferred to build the growing fetus' skeleton and used to produce milk.[16] But typically, the modest amount of bone density loss, less than 5 percent, is not a problem. Occasionally, though, women lose even more bone density during pregnancy, leading to a condition known as pregnancy-associated osteoporosis. This condition makes it more likely that a bone could be broken, most often in the spine or hip.[17]

The mortality rate for pregnant women with pelvic fractures is high—about one in ten.[18] Such a fracture could be even more dangerous for people who have lived their entire lives in a lower-gravity environment, because they have been losing bone density their entire lives. So it's likely that childbirth would be substantially more dangerous on Mars. And any risk to the mother's life puts the child's life at risk, too.[19]

That could prompt Martian births to be performed by Cesarean section, in which the baby is delivered by a surgical incision through the mother's abdomen and uterus. Because it is a surgical procedure, a C-section birth does not subject a woman to the same amount of force as a vaginal delivery. Indeed, it is possible to imagine that all births on Mars would become C-section births to maximize the safety for the mother and child. If so, that could have consequences we will consider later.

*

Other questions about human reproduction beyond Earth have to do with what happens after birth. How would a child's growth and development be affected by life in a space settlement? It is worth noting that essentially all of the information we have about how the conditions of space, like lower gravity and higher radiation, affect the body come from studies of adults. So while we do not have any data, we can at least consider what aspects of child development could be affected by the conditions kids would experience growing up on Mars.

Consider the skeleton. As children grow, their bones obviously get larger. But while some aspects of bone growth in children are determined by our genes, bone growth is also affected by how much force the bones experience. Bone tissue is constantly being replaced. Cells called osteoblasts form new bone tissue, while other cells called osteoclasts break down old bone tissue. Our bones grow when the work of the osteoblasts is greater than that of the osteoclasts. That happens throughout embryo development and childhood, and typically reaches a steady balance at some point during adolescence. But just how much bone tissue gets made depends on how hard those osteoblasts have to work during key times in a child's development. This is one of the reasons why running around and playing is such an important aspect of a healthy childhood—kids who are more active build stronger bones.

So, would growing up with one-third gravity mean that children on Mars would develop weaker bones? Quite likely. A sad but clear illustration of perhaps the closest situation we have to kids growing up in partial gravity is children who have been paralyzed or have severely limited capability for movement since birth, like some people with cerebral palsy. Given that bed rest studies have been used to simulate the effects of microgravity, people born with cerebral palsy can give us some sense of how a child's body might be affected by partial gravity. While their bodies still experience gravity, being unable to walk or, in some cases, even move their lower body reduces the amount of force imposed on their bones, especially the bones in the limbs and back. Studies of children and adults who have had this condition since birth show that they have greatly reduced bone density.[20]

If children who grew up on Mars develop weaker bones, would they be able to come to Earth? Earth's gravity would make them feel about three times heavier, and the g-forces of takeoff and entry would subject them to ever higher forces. They would likely be prone to bone fractures, particularly of bones that support our weight, like the spine and hip. Indeed, studies of people who have been paralyzed since birth show that they are at higher risk for fractures.

Another factor is the heart and other components of the cardiovascular system. Like all muscles, the strength of the heart is due in part to how hard it has to work. Heart muscles atrophy, or weaken from disuse, in adults who spend prolonged times in the weightlessness of space. It takes less force to pump blood all over the body when you don't have to work against gravity. Children might develop weaker hearts if their bodies develop in partial gravity. That could pose yet another challenge for Martian-born people traveling to Earth.

Lastly, there's the human body's largest organ—the skin. We don't often think about our skin as an organ, but it is. Anthropologist Nina Jablonski has studied human skin, its evolution, and what it does for us. "Our skin mediates the most important transactions in our lives," she wrote in *Skin: A Natural History*.[21] "Skin is key to our biology, our sensory experiences, our information gathering, and our relationships with others."

People in a Martian settlement would always be either indoors, underground, or in a spacesuit if they go outside. Some aspects of our skin adjust to our environment. Our sweat glands produce moisture to cool us down when we get hot. Tiny muscles connected to our hair follicles contract when we get cold, giving us goosebumps. Cells called melanocytes produce the pigment eumelanin in response to sunlight, which makes the skin darker and better protected from the damaging effects of ultraviolet light. How would a person whose skin had never touched the outside air or been exposed to sunlight or temperature extremes react upon experiencing these for the first time?

These are just a few of the unanswered questions about how children would be affected by growing up on Mars. Others include the social development of kids in a Martian settlement. What would a Martian childhood be like? Robert Zubrin, who has thought more about settling Mars than perhaps anyone alive today, thinks that Martian children would have to work a lot more than most kids do on Earth these days. In his book *The New World on Mars*, Zubrin wrote that "both the Martian labor shortage and the generally more serious nature of the realities of life on Mars will require that kids grow up faster." He imagines that formal education would be less emphasized than it is in most modern societies. "My guess is that, starting with elementary school, school hours on Mars will only be about half as long as is typical on Earth, with students spending half their day working and associating with adults at home, in the greenhouse, in the shop, and in the lab—getting a good hands-on education in the process," he wrote.[22]

One interesting question is how children born and raised on Mars would view their relationship to Earth and its inhabitants. Interestingly, even people who leave Earth for relatively short amounts of time sometimes return with a changed perspective. Veteran astronaut Peggy Whitson talked about this in her speech at Rice University's graduation ceremony in May 2024. She was standing in the same spot where, sixty-two years earlier, President Kennedy made his case for sending the first humans to the Moon. Whitson told the graduates seated before her about how she grew up on a farm in Iowa but dreamed of somehow going to space someday. After retiring from NASA in 2018 as the nation's most experienced astronaut, she was hired by Axiom Space as the Director of Human Spaceflight and began serving as a commander for private missions to the International Space Station—the first woman to do so.

In her remarks to the graduates, Whitson reflected on how pushing herself out of her comfort zone changed her perspective on the world—and, more specifically, on what she considered her home. "Early on, my world was family and the farm," she said. "Later that world expanded a bit to include my local school and then college. But home was still the farm. After I came to graduate school here at Rice, in an entirely different state, home became Iowa. Then I took on a job at NASA where I frequently traveled to Russia and home became the United States of America. So maybe it should not have been surprising that when I went into space, home became planet Earth."

It makes sense that people who have lived their entire lives on Earth would, upon leaving it, look back at the planet and consider it their home. The first settlers on Mars may still consider their true home to be Earth, even as they make the transition to life in their new home on the Red Planet. But what about the first generation of people born on Mars?

Immigrants to new nations often retain ties to their previous countries. The ties can be legal, if they remain citizens of their former nation. But often the stronger ties are cultural, including the language, food, and customs. These cultural identities can extend to the immigrants' children and grandchildren, but they tend to become a little weaker with each generation. My great-grandparents on my father's side immigrated to the United States from Greece and Russia. After four generations, my siblings, cousins, and I have retained only a little of the cultural identities that were so important to that first generation of immigrants.

Would the same be true of the fourth generation born and raised on Mars? Would people who had never been to Earth, and perhaps never even met a person who had been to Earth, still consider Earth their home? This seems doubtful. I think it is much more likely that they would consider themselves to be something different—they would be Martian.

But how else would the minds of Martians be modified?

5

MARTIAN MINDS

In 1961, Evgenii Shepelev sealed himself inside a small steel cylinder.

Part of a team of researchers at the Moscow Institute for Biomedical Problems, Shepelev and his colleagues were attempting to test an idea first proposed by Konstantin Tsiolkovsky, considered the father of the Russian space program, back in 1911. Tsiolkovsky surmised that deep space exploration would require a closed-loop system in which the air inside a space habitat would need to be recycled. Plants could produce the oxygen for people to breathe while absorbing the carbon dioxide they exhale. It sounds simple enough, but getting the balance just right is a matter of, well, life and death.

And now Shepelev was putting his life on the line to find out if it could work. The interior of the enclosure was only about 160 cubic feet—small enough that he would suffocate quickly unless additional oxygen was supplied. Inside the container with Shepelev was about eight gallons of water containing a type of microscopic green algae. He stayed inside for twenty-four hours—longer than would have been possible if the algae were not producing oxygen. The experiment was a proof-of-concept, demonstrating that the basic theory was sound.

But, luckily, Shepelev left the container just in time. The smell was already putrid by the time he emerged—an indication that something was off. By the next day, the algae population crashed. The Soviet researchers discovered that maintaining a balanced ecosystem inside a sealed container is complicated.[1]

In addition to the technical challenges of creating a functional, closed-loop system capable of supporting human life, there are also challenges that come with knowing you are, in fact, a human trapped inside an airtight container. Indeed, some of the most substantial challenges associated with deep space exploration and settlement are psychological.

The largest and by far the most ambitious attempt to build a closed-loop ecosystem with people inside began on September 26, 1991, when eight people stepped inside a futuristic-looking glass and steel structure in the middle of the desert in Arizona and sealed the door. They called themselves the Biospherians, and they vowed not to come out again for two years. The leaders of the project saw it as a prototype of a space community that might also provide lessons for how to better manage our ecosystems on Earth.[2] In addition to large areas for agriculture, the 7.2-million-cubic-foot enclosure contained miniature versions of five biomes—a tropical rainforest, savannah, mangrove wetland, desert, and even an ocean with a coral reef. They called the enclosure Biosphere 2—humanity's first attempt to recreate the complex ecological relationships on the surface of planet Earth, known to biologists as the Biosphere.

Each of the five biomes was packed with living things collected from equivalent ecosystems around the world. Trees, shrubs, and grasses were planted along with a menagerie of animals, including insects, fish, frogs, reptiles, birds, and mammals. "We have no way of predicting exactly which and how many of the 3,800 species Biosphere 2 contains will survive," wrote project cofounder John Allen the year the experiment began. "All that we do know is that the technology will probably keep going and the people, hopefully, will not go crazy."

It was, in hindsight, an optimistic view.

*

In May 2024 I walked through the same doorway that the Biospherians had entered at the start of their experiment thirty-three years earlier. By then the massive Biosphere 2 complex had become a research facility operated by the University of Arizona. No longer a sealed enclosure, it was now being used to study how climate change may affect the various ecosystems recreated inside. It also served as a conference venue, which is what had brought me there—to attend the annual Analog Astronaut Conference.

Analogs are a way of simulating a particular aspect of being in space. The Biosphere 2 experiment was a type of analog mission in that it was designed to simulate the experience of being confined inside a hermetically sealed habitat with a small number of people for an extended period of time. Since the early 1990s, the number and diversity of space analogs have increased substantially. I came to the conference eager to meet some of the people who have participated in these simulated missions and to get a sense for what the experiences have taught them.

Soon after arriving at Biosphere 2, I met Tim Heilers, who was attending the conference as well. He was wearing a T-shirt that read "OCCUPY MARS." We had barely gotten through the standard greetings when he asked me if I wanted to go to space.

"I guess so," I replied with a shrug. I had the feeling my answer was less enthusiastic than he expected. "What about you?" I asked.

"Oh, yes," Tim replied. "In a heartbeat." Mars, more specifically, is where he said he wants to go. "Round trip, or one-way. It doesn't matter," he added.

I asked him what it was about Mars that was so intriguing to him that he would want to leave everything behind and go there. "I've been watching *Star Trek* since I was a little kid," he explained. "I've just always wanted to go."

The conference had not yet begun, so we had some time to explore. The Biosphere 2 campus is located in the desert north of Tucson. The hulking steel and glass structure loomed in front of us with its unmistakable stepped glass pyramids and its iconic, mushroom-shaped tower. We followed the arrows painted on the sidewalk around the giant glass panels until they led to a door. We opened it, and stepped inside.

We were now in the dining room where the Biospherians had eaten their meals during the two years they were sealed inside. There was a table in the middle of the room with eight chairs. Next to the dining room, visible through an opening in one of the white walls, was a kitchen that looked like it could have been in a chic 1980s home. The combined kitchen and dining area felt spacious, with very high ceilings and floor-to-ceiling glass windows on two sides that overlooked the adjacent greenhouses. We turned to the right and followed a hallway, along which were windows where we could look into a few of the crew quarters that had been preserved as a sort of museum. Each of the Biospherians had their own two-story private suite. On the first floor was a desk and a seating area with a 1990s-era television. A spiral staircase led up to the sleeping quarters. An orange jumpsuit was hanging from the stair railing. The decor recreated its original look, with a dark patterned rug on the floor and paintings on the walls. On the desk was a notebook, opened to reveal a daily journal entry.

Seeing the personal belongings in the room made it easier to imagine what it was like to live in such a unique place. Tim and I continued down the hall and up a few stairs toward a large set of windows that looked into an enormous greenhouse shaped like a tunnel with a vaulted, curved glass roof. This had been one of the three sections of the intensive agriculture unit, where the Biospherians had grown their food. They had cultivated more than 140 different crops, including wheat, rice, corn, barley, pinto beans, squash, tomatoes, potatoes, sweet potatoes, and carrots.

"Growing enough food to feed the eight biospherians on a little over half an acre was certainly one of the greatest challenges we faced," wrote Biospherian Sally Silverstone, the food systems manager for the two-year mission.[3] "We had put together a system combining the old with the new. For example, we had incorporated an ancient Chinese method of raising fish, rice, and water fern all together with our computer-driven system for controlling temperature and humidity."

✷

As we continued our exploration of the facility, Tim seemed pensive. He told me that he has been thinking a lot about what he could do on Mars that would be useful.

"I built a Mars clock," he said. I asked what that meant. "It's a clock that tells the time on Mars," he explained. I felt silly for not having understood the simple concept after remembering that a day on Mars is forty minutes longer than on Earth. "And I figured raising insects might be good," he added.

"Have you done that?" I asked. He said that he had. He was rearing superworms—a kind of beetle larvae related to mealworms, only larger. He wasn't able to sell them, at least not yet, but he thought it would be useful experience in case he ever gets the chance to go to space.

Like Tim Heilers, the Biospherians saw their experiment as preparation for space settlement. One of them, Jane Poynter, wrote in her memoir that she had wanted to be an astronaut as a kid. "However, I didn't believe for an instant that it was a real option," she wrote.[4] But joining the crew of Biosphere 2 gave her a chance to see what life beyond Earth might be like.

"We were designing Biosphere 2 for space colonization," she wrote. "The overarching question was, is this possible? Can a life system that evolved on a planetary scale over billions of years be bottled?"

Poynter and the seven other Biospherians discovered that bottling the Earth's ecosystems is a lot more challenging than they had anticipated.

Among the first problems was a food shortage. A few months into the mission, the Biospherians discovered that a mite infestation was killing their potatoes. The entire agricultural endeavor was designed to be pesticide-free, in part because of fears that killing insects in the agricultural areas could unintentionally harm the useful insects in the other biomes. Making matters worse, Arizona had unusually cloudy weather the first year of the mission. The steel and glass structure already blocked out some sunlight, which was further diminished by the cloud cover. Short winter days made the reduction in sunlight inside the enclosure even worse. Crop yields began to decline. The crew members began losing weight. While the team's physician, Roy Walford, was an advocate of the health benefits of a low-calorie diet, even he was beginning to get concerned.

As energy levels sagged, so did morale. Inevitably, the crew started getting on each other's nerves. "We only had close physical contact with seven other people for 731 days, 2,193 meals," recalled Biospherian Mark Nelson.[5] "Pretty quickly, everyone figured out tics, habits, and how to push each other's buttons, not unlike dysfunctional family dinners. . . . Crew members deliberately provoked each other, at first earnestly and maliciously in attack and counterattack. Finally, in fun, we realized what was going on. Pushing one another's buttons was a sport that could go on endlessly—and sometimes it seemed it did—since we knew each other incredibly well."

But soon, a rift emerged that split the eight Biospherians into two equal-sized factions. Exhaustion and malnourishment had stoked the embers of a growing conflagration, but what ultimately caused it to burst into a raging inferno was a disagreement among the project's administrators.

"The core issue was whether the primary goal was to work on maintaining as self-sufficient a closure as possible and improving the facility, including trying to produce all our food," recalled Nelson. "The alternative was to lessen workloads by sending in food to give more time for research."

The project management team was led by John Allen, the visionary who had conceived of the project, and Ed Bass, the Texas billionaire who had largely financed it. Now the two had different opinions about what should be prioritized. They had appointed a Scientific Advisory Committee, which was now pushing for Allen to be replaced as the project's top research advisor. The crew members took sides. It got ugly.

"Ten months after Closure, the four of Us would huddle at one end of the dining table for lunch, and the four of Them would either leave and eat elsewhere, or crowd together at the other end of the table," Poynter wrote.

Soon, the two groups of "Us" and "Them" could no longer make eye contact. The split between the two factions would last throughout the remainder of the mission. But then a new problem added to the already strained group dynamics: oxygen levels inside the enclosure were dropping.

The enclosure was designed to have enough oxygen given off by plants and algae and consumed by people and animals, while the carbon dioxide that the people and animals produce would be taken in by the plants and algae. As Tsiolkovsky suggested, that balance would be essential for a space habitat, and no previous closed-loop experiment had ever managed to get it quite right. The percentage of oxygen in Biosphere 2 dropped from 20.9 percent to 17.4 percent in the first six months—a level that was much lower than expected. If the same rate of oxygen loss continued, the Biospherians would soon be in serious trouble.

The odd thing was that the carbon dioxide levels weren't as high as they should have been for such low oxygen levels. In a closed system, the two are tightly linked—as one goes up, the other goes down. At least it should. The search for the missing carbon dioxide had begun.

In the meantime, the oxygen levels continued to drop, reaching as low as 14.5 percent, and the crew members began to suffer. Some had sleep apnea, in which people briefly stop breathing as they sleep. Several had depression, which can be triggered by hypoxia. Roy Walford used the opportunity to capture the biological response to the declining oxygen by taking regular blood samples from crew members. He noted an effect on his own cognitive abilities. When he found he could no longer do basic arithmetic, he had to relinquish his role as lead medical officer.

*

Eventually, the missing carbon dioxide was found—it was hiding in the basement.

The concrete floor, walls, and columns of the lower levels of the structure where all the mechanical components were housed were absorbing large quantities of carbon dioxide. Too much, in fact. Some carbon dioxide was expected to be absorbed by the untreated concrete, but the quantities they found in the concrete sample cores were far beyond anticipated levels. Now the question was, where was all the excess carbon dioxide in the concrete coming from?

The culprit turned out to be bacteria. To build their biomes and farms, the project managers had included soil dredged from a local pond combined with organic peat and compost. The result was 30,000 tons of soil that was rich in nutrients as well as microbes. That was helpful for jump-starting the nutrient cycles that happen in nature, and for getting plants to grow, but the high number of bacteria had unintended consequences. The bacteria were respiring like people and animals, taking in oxygen and giving off carbon dioxide. But rather than being taken up by plants, which could make more oxygen, the excess carbon dioxide was getting soaked up by the concrete. Because carbon dioxide contains oxygen, a large amount of the oxygen in the enclosure was being taken out of circulation by the concrete.

In natural ecosystems, minerals like limestone absorb carbon dioxide. Over millions of years, the carbon and oxygen get returned to the atmosphere as these rocks undergo natural cycles of erosion, weathering, and plate tectonics. That wouldn't happen in Biosphere 2—at least not on a timescale that would be helpful to its inhabitants. While the problem could be solved for future missions simply by coating the concrete with a chemical sealant, in the short term it was an issue that needed to be dealt with.

Interestingly, Walford's blood tests showed that the crew was not adjusting to the lower oxygen levels the way people normally do when they travel to high altitudes in the mountains, where the thin air means there is less oxygen. A typical response in people coming from lower altitudes is for the body to start making more red blood cells when it senses that oxygen levels are low. For most people, it takes about five or six days for there to be enough new red blood cells to notice a difference, which is why people can usually adjust to high altitudes after about a week. Walford and his medical colleagues suggested that the crew members were not making new red blood cells in response to the low oxygen levels because their bodies didn't have enough energy to do so due to their restricted diets. They were in rough shape.

As crew health became a serious concern, the decision was made to add supplemental oxygen. The impact on the crew was immediate.

"It was a sense of well-being to breathe deeply and not feel like I needed another breath almost immediately," recalled Biospherian Linda Leigh.

"We were all madly laughing and running," remembered Mark Nelson. "Then it struck me, I haven't heard the sound of running feet in months. Friends said watching us biospherians work was like seeing a dance in slightly slow motion. With the reduced-calorie diet and lowered oxygen, people were economical in their movements. There simply wasn't extra energy to burn."

The boost in oxygen was a relief, but there was still a shortage of food. Eventually, the crew were forced to eat seeds that had been set aside for future planting, along with their emergency grain supply. This meant they were eating food that they had brought with them, not just what they grew, as was planned. Making matters worse, it threatened to diminish the amount they would be able to grow in the future. This was a further step away from their goal of creating a self-sustaining enclosure, but at least it meant they could continue to stay inside.

But different opinions about whether or not they should eat the seed reserves—and about how transparent they should be with the media about their ultimate decision to do so—drove a dagger through the heart of the project team that ultimately was irreparable. All the members of the Science Advisory Council resigned. Inside the enclosure, the brewing feud among the Biospherians reached its climax. Poynter recalled the moment when the tension boiled over:

"On Tuesday morning I was putting fodder in the animal bay when Gaie walked over to me," she wrote. "I watched as she collected a big wad of saliva in her mouth and spat in my face. She turned and walked away without a word. Forty-five minutes later I was walking up the spiral stairs during break time when Laser stopped on the stairs next to me and spat a mixture of saliva and peanut goo in my face."

Somehow, despite all the animosity the crew managed to hold out for the duration of the mission. All eight Biospherians walked out of the airlock exactly two years and twenty minutes after they had sealed themselves inside.[6] The two factions were still not on speaking terms.

"We suspected group dynamics might be one of the more challenging aspects of Biosphere 2," wrote Mark Nelson. "Turns out, we were right."

*

Tim Heilers and I continued our exploration of Biosphere 2, entering a door on the ziggurat-shaped glass structure that housed the desert biome. A boardwalk wound around the artificial rocks, where cactus, succulents, and desert shrubs grew in the sand. The walkway curved around and up a flight of stairs, and I noticed the vegetation getting thicker as the path leveled off. We reached a cliff that overlooked a miniature ocean, complete with a small beach. A solitary coconut palm tree had grown up from the sand to the top of the sloping glass wall, its fronds plastered to the glass as if trying to escape.

We stood there for a while admiring the view. Tim was quiet and had a distant look in his eyes. I asked him if he had been a part of any analog space missions. He had not—at least not yet. But he had served as a sonar technician on a nuclear submarine in the Navy. That struck me as pretty good practice for being in an isolated, confined, and hostile environment much like in space. I said so, and Tim agreed.

Compared with a crowded submarine, the Biospherians had had quite a lot of space—combining all the indoor areas, it came to more than 17,000 square feet per person. Still, it often felt confining, especially as the rift between the crew members deepened. "With nearly palpable hostility hanging in the Biosphere like a toxic cloud, Taber and I looked for ways to escape," Poynter recalled. "Sometimes we hiked the two hundred feet down the stairs and through the basement to the beach and sat looking out through the spaceframe at the stars. We would search for Mars in the night sky and contemplate how different things might be if our

Biosphere were suddenly transported there and we were looking not at the tiny red speck of Mars, but at the bluish dot of our home planet. Perhaps on Mars, with the safety of home at least forty-eight million miles away, we would have been able to pull together. But then again, perhaps not. I wondered if people are really meant to be enclosed in small spaces, even as large, beautiful, and varied as Biosphere 2. The human species, after all, did not evolve indoors."

The same question had brought me to the Analog Astronaut Conference, and I was eager to find out what lessons had been learned in the three decades since the tumultuous Biosphere 2 experiment. Tim and I made our way through the final enclosed portion—the moist, green rainforest biome—before emerging once again into the dry desert air outside. All of us conference attendees were staying adjacent to the original Biosphere facility in colorful adobe-style casitas.

I met my roommate, Donald Jacques, a sixty-six-year-old Air Force veteran and former computer programmer who now works as a handyman. He told me he was building his own space analog out of a bus that he has lived in full-time for over a year. He calls it the MASH, short for Mobile Analog Sustainability Habitat. Donald eagerly explained how he uses black soldier fly larvae to break down solid human waste and convert it into soil. The extra flies and larvae get fed to his chickens. The chicken waste, in return, gets broken down by the fly larvae. His goal, he said, is to create a sustainable, self-enclosed system as a prototype for a space settlement. I was impressed by the effort and dedication. It sounded like a mini-Biosphere 2 on wheels.

"What's your motivation?" I asked him.

Donald told me that he wants to visit the Moon, then Mars, which he thinks will be possible soon. From Mars, he wants to travel to places beyond, like the moons of Jupiter and Saturn, ultimately to destinations outside our solar system. He sees the work he is doing on the mobile habitat as his ticket to the stars.

*

The idea that participating in analog simulations can prepare a person for actual spaceflight is embodied by the Analog Astronaut Conference's matriarch and cofounder, Sian Proctor. Like many in the analog astronaut community, Proctor had been trying to get to space her entire life. She was born on Guam, where her father was working at a tracking station as part of the Apollo program.

I had seen her speak at the Humans to Mars Summit in DC the year before. "I'm a Moon celebration baby," she had said. "I was born eight and half months after the first Apollo landing."

Growing up around the space industry in the 1970s and 1980s, Proctor dreamed of becoming an astronaut herself. "As a kid, I didn't realize there weren't yet any black astronauts," she said. So, in her thirties she applied to be a NASA astronaut, and in 2019 she made it all the way through the selection process until the final step.

"It came down to the yes/no phone call," she said.

The call came, but it was a no. Disappointed but undeterred, Proctor continued to pursue her passion through analog space missions. The following year she joined a two-week analog mission at HI-SEAS, a facility located at 8,200 feet above sea level on the slopes of Mauna Loa. She had done her first analog mission there—in fact, she was part of the first ever HI-SEAS analog simulation in 2013. She also did two-week simulations at the Mars Desert Research Station in Utah and at another analog facility in Poland.

Then, one day, she got the call she had been waiting for her entire life. It was Jared Isaacman, an entrepreneur financing a mission to space called Inspiration4 that was slated to fly a crew of four people on a SpaceX rocket. It would be the first-ever all-civilian mission to orbit the Earth. He invited Sian Proctor to join him.

The Inspiration4 crew orbited the Earth for four days in 2021, with Proctor as its pilot. That made her the first female commercial astronaut pilot and the first African American woman to be a pilot on any spacecraft. She became something of a celebrity in the space community, and the role model for every analog astronaut hoping that their dreams—like Sian Proctor's—will somehow come true.

I asked Proctor if the analog missions had been helpful preparation for the real thing. "Oh yes," she told me. "They taught me important lessons about myself and my capabilities, as well as my limitations. And they taught me about how to work well in a team," she said.

NASA sees analogs as useful as well.[7] It operates several of its own, including two at Johnson Space Center. HERA, the Human Exploration Research Analog, is a 650-square-foot habitat where crews of four spend up to forty-five days on simulated space missions. The 1,700-square-foot CHAPEA analog, which was 3D-printed to simulate a habitat on Mars, completed its first year-long mission with a team of four in July 2024. NASA also operates an underwater analog in the Florida Keys, called the NASA Extreme Environment Mission Operations project, or NEEMO. Other governmental space agencies operate their own analogs, including a series of hermetically sealed habitats at the Institute of Biomedical Problems in Moscow that has been used by the Russian space agency, ROSCOSMOS, as well as the European Space Agency.

A clear advantage of analog studies is that analogs are much less expensive than going to space. That means that a lot more analog studies can be conducted and that they are accessible to a wider range of people. Some analog facilities can be modified to test the effects of particular aspects of a space mission. For example, a communication delay of around twenty minutes can be implemented to see what the effect is on the crew of having a delay similar to that between Earth and Mars.

Psychologist Nick Kanas has summarized the sources of psychological stress that come with prolonged spaceflight that can be examined using analog missions.[8] They include the stress of isolation and confinement, boredom from having too much free time and too little stimulation, repetitive work and monotonous days, and conflicts and disputes with fellow crew members. Some interesting results of these studies include findings related to the size and composition of the crew. In general, the larger the crew size, the less likely that any one crew member feels isolated. An odd number of crew members makes it easier for the group to come to agreements, since it's impossible to split an odd number into two evenly sized factions, as happened to the Biospherians.

*

One type of stress that analogs can't always simulate is the life-threatening danger from the extreme environments of space. If something goes seriously wrong in a simulated mission, the crew can simply leave. Simulations are certainly safer for the crew than being in an environment that more accurately recreates the harshness and isolation of being on the Moon or Mars. But, on the other hand, the fact that the crew know they can leave if there's an emergency must have an effect as well.

Angelo Vermeulen, a Belgian biologist, artist, and engineer, served as commander of the first analog mission at HI-SEAS along with Sian Proctor. He told me that he had a hard time imagining he was on another planet when he was very well aware that he was in fact in Hawaii.

"Just looking outside and seeing a blue sky, and a bird passing by—it's not Mars," he said. "It's difficult to maintain the belief you're pretending to be on Mars for such a long time." Although the location of HI-SEAS is relatively remote, in an emergency it would be possible to just step outside the facility. Even though the crew wore simulated spacesuits to keep up the feeling that they were on Mars, at the end of the day they all knew the suits weren't actually necessary.

"I think it's different when you have an underwater station," he offered. "Probably that is a little closer to feeling like you're surrounded by a threatening and hostile environment. Of course, if you're in Antarctica . . . that's a different story. Then you're really in a hostile environment and also very remote where there is no help available. I think that would psychologically be very different."

Indeed, studies of people living and working at polar outposts have provided some of the best analogs for deep space missions. Antarctica in particular serves as the place on Earth that most closely resembles the extreme conditions on the Moon or Mars. While there are no permanent settlements on the white continent, there are dozens of permanent research bases. Some, like Palmer Station, can be accessed year-round by ship. Many others, however, are only reachable during the summer season. The largest station in Antarctica, McMurdo, houses some 1,500 people in the summer when it can be reached by ship. It also has landing strips for airplanes and helicopters, but there are no real transportation options during the winter when it's nearly always dark and temperatures average around –18 degrees Fahrenheit with plenty of snow. The roughly 200 researchers and staff who spend the entire winter at McMurdo experience true isolation in the most extreme conditions.[9]

Seasonal affective disorder is common at McMurdo, afflicting more than a quarter of those who overwinter. Some develop a more extreme version known as "winter-over" syndrome that includes having a hard time focusing and can include becoming irritable or even aggressive.[10] Some develop a blank look known as the "Antarctic stare." About one out of every twenty people who overwinter in Antarctica develop true psychiatric disorders.[11]

Another group of people who experience both the isolation and confinement of space with a truly extreme and hostile environment are those who serve on nuclear submarines, like Tim Heilers. Nuclear submarines can remain underwater for months at a time, and being submerged in the deep sea certainly qualifies as being surrounded by a hazardous environment. While some studies have been conducted on the group dynamics and psychology of submariners, their being part of military operations makes it difficult for researchers to have the same kind of access as in many space analogs.

As extreme as the Antarctic bases and nuclear submarines are, even these lack some of the conditions that make space so hazardous, like reduced gravity and high radiation. On the other hand, that allows researchers to study the effects of individual stressors, such as being in a confined environment or having limited social interactions, which could lead to finding ways to help mitigate them.

One of the most salient lessons from the world of analogs seems to be that regardless of whether they're in a submarine, a polar research station, or a simulated Martian habitat, it makes a big difference whom you seal up inside a small space for a prolonged time. Psychological screening, which can include tests, interviews, and a review of personal and family history, can minimize the chances that a person more likely to develop mental health issues gets included in a crew. But in addition to selecting *out* people with certain characteristics that might become problematic, selecting *in* people with beneficial personality traits can also be helpful. In general, people who are highly motivated, deal with stress in healthy ways, and work well with others tend to do well in both analog and real space missions.

But individual traits alone are only part of the recipe; it also matters how they work together as a team. It is difficult to know how well any group would do until they are actually faced with challenging conditions. For this reason, training and group activities in advance of a mission are often used to evaluate candidates and build team cohesion.

For example, Angelo Vermeulen told me that prior to their four-month HI-SEAS analog mission, he and his crew members spent two weeks training at the Mars Desert Research Station in Utah. "Interestingly, the rules, the social dynamics—everything that we used in HI-SEAS—was developed in Utah," he said. "You come together as a group, you establish a kind of relationship with each other and you don't really change that anymore. It's a typical group dynamic that sets in pretty quickly."

Still, there is no guarantee that even a carefully selected and trained crew is going to work well together. The Biosphere 2 crew had undergone quite a bit of training and work together before beginning their two-year stint inside the sealed facility in Arizona. But that didn't prevent the team of eight from deteriorating into a dysfunctional pair of factions who were more likely to spit on one another than to have a conversation.

At one point during her time inside Biosphere 2, in the midst of all the hunger, fatigue, and frustration, Jane Poynter recalls a moment of peace and clarity while pulling out weeds in the desert biome. The idea that she controlled which plants lived and which one died struck her as an awesome, almost Godlike, power.

"I wondered whether Biosphere 2 was an attempt at the ultimate expression of control—the bottling of nature to thoroughly conquer it once and for all," she mused.[12] "If we can cage nature, tame it, and train it to work for us, then we have overcome a distinct glitch in our quest for species immortality. After all, every living thing's primary innate goal is to have offspring to perpetuate the species. Biospheres simply promise to take that one step further: the species can ensure that it has an environment fit to inhabit."

The idea made enclosures like the one she was locked inside suddenly seem like a very real possibility for future space settlements. "At that moment, fantasies of biospheres floating through space, keeping cities of humans alive, seemed perfectly plausible," she recalled. "Whether future space was to be populated with sterile, overcrowded tin cans or biospheres, I only hoped that the people and their cultures would have evolved to get along better than we had."

*

One psychological aspect of real spaceflight that is not replicated by any analog simulations is the profound impact of seeing Earth from space.

"It's hard to describe the experience of looking down at the planet. I feel as if I know the Earth in an intimate way most people don't," wrote astronaut Scott Kelly after his year aboard the International Space Station.[13] "One of my favorite views of the Earth is of the Bahamas—a large archipelago with a stunning contrast from light to dark colors. The vibrant blue of the ocean mixes with a much brighter turquoise, swirled with something almost like gold, where the sun bounces off the sandy shallows and reefs."

Unlike in a ground-based analog, the awe and wonder of being in space and seeing the Earth from the outside can help to offset some of the challenging aspects of being confined and isolated with a small group of other people. "Sometimes when I'm looking out the window it occurs to me that everything that matters to me, every person who has ever lived and died (besides the six of us [on the International Space Station]), is down there," Kelly wrote. "Other times, of course, I'm aware that the people on the space station with me are the whole of humanity for me now. If I'm going to talk to someone in the flesh, look someone in the eye, ask someone for help, share a meal with someone, it will be one of them."

In the 1980s, space philosopher Frank White created a name for this shift in perspective that astronauts often report having while in space—he called it the "overview effect." The idea came to White while flying on an airplane. "As the plane flew north of Washington, D.C., I found myself looking down at the Capitol and Washington Monument," he wrote in his 1987 book about the idea.[14] "From 30,000 feet, they looked like little toys sparkling in the sunshine. From that altitude, all of Washington looked small and insignificant." Looking out over the landscape also gave him a sense of how interconnected everything is. "From the airplane, the message that scientists, philosophers, spiritual teachers, and systems theorists have been trying to tell us for centuries was obvious: everything was interconnected and interrelated, each part of a subsystem of a larger whole system," he wrote.

White wondered whether being even higher would create an even more profound effect. He interviewed astronauts who had been in Earth orbit or to the Moon, and discovered that many of them said they had indeed experienced a major shift in perspective. White discovered a talk given by Apollo 9 astronaut Rusty Schweickart in which he described his personal experience orbiting the Earth in March 1969: "When you go around the Earth in an hour and a half, you begin to recognize that your identity is with that whole thing. That makes a change. You look down there and you can't imagine how many borders and boundaries you cross . . . you don't even see them . . . from where you see it, the thing is a whole, and it's so beautiful," he said.

Seeing this view of the world quite literally changed Schweickart's worldview.

"You look down and see the surface of that globe that you've lived on all this time, and you know all those people down there and they are like you—they are you—and somehow you represent them. You are up there as the sensing element, that point out on the end, and that's a humbling feeling. . . . And somehow you recognize that you're a piece of this total life. . . . That becomes a special responsibility and it tells you something about your relationship with this thing we call life," he said.

White later interviewed Schweickart, who expanded on his previous insights. Other astronauts told White similar things. Many said they became more interested in protecting the fragile planet we live on after seeing it from outer space.

"Some parts of the world . . . are so blanketed by air pollution that they appear sick, in need of treatment or at least a chance to heal," wrote Scott Kelly. "The line of our atmosphere on the horizon looks as thin as a contact lens over an eye, and its fragility seems to demand our attention."

Hearing from astronauts convinced Frank White that he was onto something. While images from space could have an effect on people—the Earthrise photo taken by Apollo 8 astronaut Bill Anders while orbiting the Moon helped to launch the environmental movement[15]—experiencing the view personally seemed to be much more impactful. White thought about the long-term consequences of the overview effect, not only for individuals but for all of humanity. If each person who goes to space comes back transformed, he reasoned, imagine what would happen once many people have been to space.

And imagine what would happen to people who leave our planet to live off-world permanently—and what would happen to their children.

I met Frank White at a space conference in Austin called New Worlds. I sensed he had become a sort of patriarch of the space community. It was clear that he was widely respected among the conference attendees, who spoke of the overview effect with the sort of reverence normally associated with religious doctrine. We spoke again by Zoom a few days later once he was back at his home in Massachusetts. Given how much of White's work focused on the psychological effects of actually being in space, I wondered what he thought about analog missions.

"I think the people who are doing the analogs are the best candidates for large-scale space migration," he said. "They're not focusing on 'I want to get on a rocket.' They're focusing on 'I want to know how to function in a small team out there once I get my chance.' And I think that's wise."

The conversation soon turned to the topic I most wanted to ask him about, something we had both thought deeply about: the long-term consequences of space settlement. White said he considered the topic so important that he made it the subtitle of his book—*The Overview Effect: Space Exploration and Human Evolution*.

"The more I worked on the book, the more evolution became important," he told me. White was quick to point out that he is not a biologist, but based on his understanding of evolution he had come to some conclusions about how space exploration and migration could be a catalyst for changes that are both cultural and biological. "It seemed really, really clear that space exploration would trigger evolution," he said.

I nodded. The question, we both agreed, was *how* people in space would evolve.

6

COSMIC ISLANDS

At first, Charles Darwin didn't think much of the Galápagos Islands.

"These islands at a distance have a sloping uniform outline," he recorded in his journal upon the HMS *Beagle*'s arrival in September 1835.[1] "The whole is black lava, completely covered by small leafless brushwood & low trees . . . the stunted trees show little signs of life. . . . The plants also smell unpleasantly. The country was compared to what we might imagine the cultivated parts of the Infernal regions to be." It was not a great first impression.

What Darwin saw there would eventually lead him toward a revolutionary theory about how species change, a process we must understand if we want to predict how our descendants on Mars—or anywhere—many generations in the future will be different from us. But when Darwin first set foot in Galápagos he was not yet much of a biologist. As a recent university graduate with a degree in theology and a fondness for natural history, particularly geology, he was looking forward to visiting Galápagos primarily for the chance to see active volcanoes.

He was modestly interested in the animals, as well, writing in a letter that "both the geology & zoology cannot fail to be interesting." Yet his first impression of the fauna was just as gloomy as his descriptions of the landscape. Seeing marine iguanas on the shoreline, he described them as "most disgusting, clumsy lizards." The flora didn't fare much better in his eyes; the first day he collected ten types of what he called "insignificant, ugly little flowers."

The *Beagle* only spent five weeks there, but a few key observations from Galápagos would prove to be pivotal for Darwin.[2] A key moment came when he was told by the vice governor that it was possible to tell which island a tortoise came from based on the shape of its shell. Darwin had by then visited several islands, and he thought they all seemed quite similar. If the environment was identical, and the islands were all close together, why would a tortoise from one island be any different than a tortoise from another? Come to think of it, he had noted that the mockingbirds on different islands had slightly different beaks. Until then, he had not bothered to record the island on which each of his specimens was collected, thinking it was an irrelevant detail.

He also noted that the Galápagos fauna seemed closely allied with the animals from South America, where the *Beagle* had been for the last three and a half years. The marine iguanas and land iguanas, both of which are only found in Galápagos, reminded him of South American iguanas. The mockingbirds were similar to South American species, too. "It will be very interesting," he wrote in this journal, "to find from future comparison to what district or 'centre of creation' the organized beings of this archipelago must be attached." It was not until he returned to England, had given his collections to experts to examine, and had time to reflect on it all that his observations and ideas finally coalesced into a single, unmistakable conclusion: species change.

That part was shocking enough, particularly to the Christian worldview that dominated Western civilization. But Darwin's key insight was figuring out *how* they change. After his return to England in 1836, he spent the next two decades devising and then refining a theory of evolution and accumulating evidence to support it. It was finally published on November 22, 1859, as *On the Origin of Species by Means of Natural Selection, or The Preservation of Favoured Races in the Struggle for Life.*[3] The title would soon be shortened to *On the Origin of Species*.

The basic thesis was that a process he dubbed "natural selection" was the mechanism by which species become adapted to their environments. As the environment changes, individuals with traits that are better for survival will have more chances to reproduce, passing their traits to their offspring. Over generations those traits will become more common. The ancestors of Galápagos marine iguanas couldn't swim in the ocean, but over generations those that were able to spend a little more time underwater were able to eat more algae, and that gave them a survival advantage. Adaptation was the outcome, and natural selection operating over generations was the process that made it happen.

Darwin concluded the book by pointing out not only that species *evolved*, but also that they are *evolving*. In other words, evolution is a process—one that we shouldn't expect to ever be complete.

*

If evolution is constantly happening, as Darwin suggested, does that mean humans are still evolving?

Indeed, we are. While many people have access to abundant food, clean water, and advanced healthcare and medicine, not everyone does. Unfortunately, infectious disease is still a leading cause of death in much of the world. In regions where diseases like malaria are prevalent but treatment is not, anyone who is genetically blessed with natural resistance is more likely to pass their genes to the next generation than someone who is not. And even in regions where healthcare is more accessible, there are still differences in life expectancy that are rooted in the genetic cards we are each dealt at birth.[4]

The COVID-19 pandemic was a stark reminder of this. Before vaccines and treatments became available, some people who contracted the SARS-CoV-2 coronavirus became extremely ill and, in some cases, died, while others were less severely affected. Research has found that these differences were primarily due to genetics, in which some people are naturally more vulnerable to the virus than others.[5] Those fortunate enough to have genes that made them less vulnerable were more likely to survive.

Natural selection is a cruel process, but it works. And it never stops working.

We've now assembled all the pieces we need in order to put together a coherent vision of the future of our species beyond Earth. Understanding *how* natural selection operates allows us to make informed predictions about how people would adapt biologically to a particular environment. We have a good idea of what the environment is going to be like on Mars thanks to the rovers and other devices that have been sent there to make detailed measurements and observations. And we know quite a bit about how individual people are affected by those conditions, like lower gravity and higher radiation, thanks to decades of studies on space medicine.

So, putting it all together, what types of biological adaptations might we see in future generations of settlers on Mars?

Let's start with gravity. We know that adults who travel to space and experience weightlessness grow a little bit taller as the discs that separate the vertebrae in their spines become less compressed and their spines straighten. But these changes are not heritable—the children of astronauts are not taller than their parents. Or, if they are, it has nothing to do with their parents having spent time in space. Still, it is possible that future generations of Martians could be taller. With three-eighths the gravity of Earth, a person on Mars would weigh three-eighths as much as a person with the same mass on Earth. Without having to support as much weight, over generations Martians might evolve a spine with less curvature, which would make them a little taller.

On the other hand, applying Darwinian thinking suggests the opposite may be true. We have seen how our bones grow in response to the forces being imposed on them, so in a three-eighths-g environment people would lose bone density as they age. If C-sections do not become the standard way for Martian babies to be delivered, then childbirth could become much riskier. If we assume that all people lose bone density at about the same rate, then women who happen to be born with slightly denser bones could have a lower risk of a fracture during childbirth because their bones would be stronger. These women would be more likely to survive and to pass on their genes for dense bones to their children. In other words, natural selection would favor denser bones.

Over many generations, this could lead to Martians having much more robust skeletons. In that sense their bones could resemble those of some early human ancestors. And the easiest way to have denser bones is for the body to make bones with the same amount of bone tissue but to make the bones smaller. In other words, for people to be shorter.

On the other hand, if C-section births do become ubiquitous on Mars to protect the lives of brittle-boned mothers and their babies, that could lead to some other scenarios. Throughout our evolutionary history, there has been a constraint on the size of the head because it must pass through the birth canal. If all babies were delivered surgically, there would no longer be an upper limit on head size, other than the fact that our necks still have to support them. In theory, this could allow our heads to expand. If natural selection favored larger heads for other reasons, perhaps because of an increase in intelligence, then Martians could evolve to look something more like the stereotypical big-headed aliens from science fiction.

There is also a potential downside to all births being done by C-section—we could become dependent on them. If infant head sizes become so large that they cannot possibly fit through the birth canal, then our species would require surgical births just to perpetuate itself.[6] If for any reason the ability to perform such surgeries were compromised—by, say, a lack of suitable materials for surgical instruments or inadequate facilities for performing an operation using sterile techniques—the result would be almost certain extinction. Biologists call this type of thing, in which natural selection favors a trait or behavior that ultimately becomes self-destructive, an evolutionary trap.[7]

*

Being in a lower-gravity environment could also change other aspects of our anatomy. The fossil record shows that as brain size increased in early species of our genus, *Homo*, the shape of the pelvis changed along with it. Even before our brains were getting bigger, the pelvis had already changed quite a bit in shape during the earlier transition to walking upright. In a four-legged animal the pelvis is primarily an attachment point for the hind legs. But walking upright on two legs requires the pelvis to support the weight of the entire upper body, which meant that the pelvis had to become wider and more bowl-shaped. As our heads got larger, the pelvis had to keep supporting the upper body while expanding the size of the birth canal.[8] But those two functions are somewhat at odds—the pelvic ring that forms the birth canal can grow only so large before it compromises the ability of the pelvis to support the upper body.

At least that is the case in Earth's gravity. In a lower-gravity environment, like that of Mars, the pelvic ring might evolve to be larger because there would be less force pushing down on it because the upper body would weigh less. So, Martians could evolve wider hips, which in turn could allow the evolution of babies with larger heads even without having all C-section births.

The fluid shifts that plague astronauts, giving them puffy faces and chicken legs, come about because the body has developed an amount of fluid appropriate for Earth's gravity, but that turns out to be too much for microgravity. Fluid shifts are thought to cause problems with their eyes like the choroidal folds that Scott Kelly experienced. That means that people born in a lower-gravity environment who naturally produce less body fluid would be less likely to develop vision problems. Over generations natural selection could favor people with increasingly less body fluid. That would further lower their body weight, while avoiding having skinny-legged Martians with puffy faces.

The arches in our feet could become less pronounced in lower gravity as well. Normally, children develop arches in their feet by the age of six, which help put a literal spring in our step. However, between one-fifth and one-third of the population never develops them, a condition known as being flat-footed. Being flat-footed runs in families, suggesting it has a genetic basis. That means it could evolve. While being flat-footed can be a painful condition here on Earth, it would be less problematic on Mars because they would be putting only about a third of the weight on their feet. While it might not be beneficial to be flat-footed on Mars, it also might not be harmful, meaning it might not be selected against. Evolutionary biologists call this a relaxation of natural selection. Over generations, relaxed selection for foot arches could make for more flat-footed Martians.

By the same reasoning, the fact that people on Mars would likely spend all of their time in either a climate-controlled habitat or a spacesuit could lead to relaxation of selection for traits like our ability to sweat. Our ancestors evolved a much larger number of sweat glands as an adaptation to help them cool off while chasing down animals on the African savanna.[9] Without a need to cool down, the number of sweat glands could decline among Martians. That could change the way they smell by reducing the intensity of their body odor.

*

Living underground would also remove people from the natural daily and annual rhythms we often take for granted. Our bodies naturally produce hormones at different levels depending on the time of day.[10] In the morning, hormones like cortisol signal the body that it's time to wake up and begin the day. In the evening, the production of melatonin helps us fall asleep. The timing of these hormone releases evolved in our ancestors based on cues from the environment, namely, sunrise and sunset. The timing is somewhat flexible, as we experience during seasonal changes in day length and when we travel to another time zone.

A sol on Mars is forty minutes longer than a day on Earth, as Tim Heilers's clock shows. That difference may seem small, but it's enough to cause sleep disruptions among scientists on Earth working on a Martian sol schedule, as Kirsten Siebach told me. While the timing of day/night cycles in an underground habitat could be controlled and kept synchronized with Earth, that would likely be disruptive for Martians because their sleep cycles would not correspond to the cycles of light and darkness on their planet. Instead, I think it is likely that Martians would keep their own schedules. If so, then after several generations the daily rhythms driven by the timing of hormone releases could evolve to match the length of a Martian sol.

Annual cycles would be even more different on Mars, since a year there is nearly twice as long as a year on Earth. Much as with daily cycles, our bodies produce different levels of hormones depending on the time of year. For example, thyroid-stimulating hormone, which helps regulate metabolism, is often higher in the summer. Living underground could help insulate people from seasonal changes, although again I think people would want to stay connected to what is happening at the surface. The extent of seasonal change would also depend on whether settlements are built closer to the equator or the poles. The equator might be a good choice, because people who suffer from seasonal affective disorder might find the long Martian winters unbearable. Natural selection could adapt people to the Martian seasonal cycle by changing the timing of their hormone production to match that of their new planetary home.

The content of the Martian atmosphere will be a challenge, although the MOXIE experiment on the Perseverance rover showed that it is possible to produce breathable oxygen from the carbon dioxide in the air. Oxygen could still be a limited resource on Mars, though, due to the planet's very thin atmosphere. But we know that people who have lived for many generations in high-altitude areas have evolved ways of compensating for the low levels of oxygen in the thin mountain air.[11] Martians may evolve similar adaptations for tolerating hypoxia, such as an expanded lung capacity or denser capillaries.

We could also see Martians adapting to the perchlorates in Martian soil. While most organisms find perchlorates toxic, some microorganisms on Earth have evolved to be able to survive in areas with high levels of perchlorate.[12] Any slight difference among individuals in their tolerance of perchlorates could, over many generations, allow natural selection to improve perchlorate tolerance among the entire Martian population.

Radiation is another major way in which the environment on Mars differs from what our species evolved to deal with on Earth. Based on Catherine Davis-Takacs' work on rats at Brookhaven National Lab, the radiation exposure for people on Mars could cause cognitive effects like delays in reaction time. Any individuals who were slightly less susceptible to these effects might have a survival advantage. If so, natural selection could lead to cognitive improvements in future generations of Martians. That wouldn't necessarily make them smarter, but it could at least help them avoid losing their edge—and could give long-term Martian residents an advantage over new arrivals.

Most proposals for habitats on Mars include some type of radiation shielding, including living underground. Yet I believe people would refuse to spend all of their time inside, no matter how attractive the architecture. Any amount of time spent on the surface, even with a spacesuit on, would increase a person's lifetime radiation dose. Cancer rates would be high on Mars, especially in the first few generations. However, natural selection would favor any genes that give a person a better ability either to prevent DNA damage caused by radiation or to repair it.[13] Some people's bodies are already naturally better at this than others, again thanks to the genetic lottery. For example, people who produce more of the skin pigment eumelanin are better protected from DNA damage caused by ultraviolet radiation than those with less. For this reason, darker skin colors might be favored by natural selection for people on Mars.

Another possibility is the evolution of new types of skin pigments to protect Martians from radiation. For example, carotenoids are orange or yellow pigments commonly found in plants and fungi that play a role in radiation protection. We get carotenoids by eating food that contains them, like carrots. But some insects have acquired the genes to make carotenoids themselves, suggesting a similar thing could happen among space settlers—perhaps giving rise to Martians with orange skin. Various other types of genes, like the kind known as tumor suppressor genes, also help to detect DNA damage when it does occur and to destroy cells that could become cancerous.[14] Those genes could also become more common as people become better adapted to life on Mars.

One option would be to screen for people with these genetic advantages and choose them to be the ones who go to Mars. Another option, which we will explore later, is to modify people's genes to improve their cancer-fighting abilities. But if we choose not to do either of these, then people who happen to have traits that help them survive a little longer would have more opportunities to pass their genes to later generations. As Darwin figured out from his journey on the *Beagle*, that means they would adapt.

But it doesn't mean it would be a pleasant process. Based on his observations in Galápagos, Darwin wrote about what he called the "struggle for existence." For at least the first few generations of people on Mars, "struggle" will be an understatement.

*

It is worth noting one other important aspect of adaptation that Darwin figured out. One of his greatest insights was to recognize that the variation that exists in a species is the raw material available for natural selection to operate. Darwin wasn't quite sure where that variation came from—our understanding of genetics would not mature until the twentieth century—but he could tell that heritable traits exist, and for natural selection to work there had to be some difference in them from one individual to another. And the more variation there is, the more potential exists for natural selection to create adaptation. That means that for people to adapt to the conditions on Mars, the more variation there is in the population, the better. Sending a diverse starting population would be wise, including for other reasons we will soon examine.

But we should also consider the ultimate source of all variation: mutations.

While some mutations don't have any effect, most mutations are harmful—like those that lead to cancer. However, on occasion a mutation happens that actually has a beneficial effect. Mutations occur randomly, but the more mutations there are, the greater the odds that one turns up that happens to be useful. Because radiation exposures would likely be higher on Mars, the mutation rate would be higher as well. That would create a larger pool of genetic variation for natural selection. In that sense, radiation could jump-start the adaptation process.

We tend to think of evolutionary changes as being very slow and gradual. Darwin certainly did—he used the word "slow" 144 times in *On the Origin of Species*. But we have learned since his lifetime that evolutionary change doesn't always happen slowly. In some cases, we see it happening much faster than we would like, such as when bacteria evolve resistance to antibiotics. What evolutionary biologists have figured out is that there are three main factors that determine how quickly a species evolves.[15] The first is one we have just considered—how much genetic variation exists within the species. The second is the strength of natural selection, meaning how much of an advantage a particular trait gives an individual when it comes to survival and reproduction. The third is the generation time, in other words how quickly the species reproduces. Bacteria evolve extremely fast in part because some species can make a copy of themselves as quickly as every twenty minutes.

But even species that reproduce slowly can evolve rapidly if there is a sufficient amount of genetic variation and strong enough selection pressure. Consider the case of elephants in the nation of Mozambique, where civil war between 1977 and 1992 contributed to very high rates of poaching. Poachers were killing elephants primarily to sell their tusks in the ivory trade. Some female elephants are naturally born without tusks, and these elephants were five times more likely to survive during the civil war than those with tusks. So, there was variation in the trait—some elephants had tusks and some didn't—and strong natural selection favoring tuskless individuals. Tusklessness is a heritable trait, meaning that tuskless mothers are more likely to have tuskless daughters (due to a genetic quirk, males cannot be tuskless). The result was a threefold increase in the number of tuskless elephants, from about one in every five to roughly half the female population in just twenty-eight years.[16] That's pretty fast considering the average generation time for elephants is about twenty-five years—roughly the same as for humans.

What this means for humans living on Mars is that we should not assume that their evolution would happen slowly.

Rather, the ingredients for rapid adaptation will all be there. Genetic variation will exist as a result of both the people we choose to send and the accumulation of new mutations from radiation exposure. And the harsh conditions on Mars, including much higher radiation and lower gravity than our species has experienced during its evolutionary history, will translate into very strong natural selection. I estimate we could see noticeable evolutionary changes after as little as four or five generations, with more significant changes occurring after ten or more. We're still talking hundreds of years, which may seem slow in the context of a human lifetime. But for evolution, that's the blink of an eye. If Charles Darwin were around to see it, sailing between the planets on a space-age version of the *Beagle,* I imagine he would recognize the process but might raise an eyebrow at the speed.

But there are other factors we should consider that will influence in what ways and how quickly humans will evolve beyond Earth. For that, we need to do some more island hopping.

*

In 1929, a seventy-five-ton schooner named *France* was making its way through the tropical islands north of Australia. The team was nine years into the journey, known as the Whitney South Seas Expedition. Funded by American businessman Harry Payne Whitney, the main goal was to collect birds for the American Museum of Natural History in New York. They had already visited more than a hundred islands, from the far-flung atolls of eastern Polynesia to the rugged, jungle-clad slopes of New Guinea. But now the expedition leader and the team's main ornithologist, Rollo Beck, had left, and they needed someone knowledgeable about birds to keep the expedition going.

The news reached a young German biologist named Ernst Mayr, who happened to be in the midst of his own ornithological expedition in New Guinea. Mayr was skinning birds in a small hut when a messenger delivered a telegram. It was from his mentor, Erwin Stresemann, who informed Mayr about the vacancy in the Whitney expedition and encouraged him to take the job. Doing so would be helpful to Mayr's career, his mentor was suggesting, and besides, Stresemann was promised some of the specimens if Mayr agreed.

"Whether I liked it or not I had to say yes," Mayr later recalled.[17] He left his camp and made his way to the coast, then caught a steamer to the eastern tip of New Guinea where he met up with the remaining members of the Whitney expedition aboard the *France*.

"Everybody thinks what a wonderful thing being on [a] schooner cruising around in the South Islands [is]," Mayr remembered. "Well, this was an old copra-carrying freight schooner, most uncomfortable. It had no actually working toilet facility. Under deck if you tried to sleep in the bunks, it was always much too hot, you couldn't fall asleep so you slept on top of the deck, and, of course, every night a rain shower came sooner or later and drenched you."

Despite the discomforts, Mayr spent nine months aboard the *France* collecting birds in the Solomon Islands to the east of New Guinea. They visited several places, including the island of Malaita and the highlands of San Cristobal, where no previous scientific expeditions had ever collected birds. Mayr described the collections as "ornithologically very successful." While the Whitney expedition would continue after Mayr returned home, his relatively short stint on it would later prove to be pivotal for his career. Soon after returning home, Mayr was offered a temporary position at the American Museum of Natural History to curate the specimens from the Whitney expedition. The position was extended for a second year, and then, eventually, he was offered a permanent curator position.

As he worked his way through the extensive collections of birds from across the South Pacific, Mayr began to notice some patterns. He saw that while some species were widespread, others were completely restricted to particular islands or groups of islands. Mayr was well aware of Darwin's work on natural selection leading to evolutionary changes within a species. That part made sense to him, and he assumed it applied to his birds. But the patterns he saw in the distribution of birds on the Pacific Islands were suggesting something more. Thinking about how the birds were all related, he could imagine them all sharing a common ancestor that somehow wound up spread across multiple islands. The fact that they were on different islands seemed to be a key component that led them to evolve into different species.

Until then, the concept of species was somewhat loosely defined by biologists. Darwin wrote about the origin of species, but he didn't actually define what the term "species" meant. In some cases, it seemed fairly obvious. Horses and zebras are different species because they have obvious features that make it easy to tell them apart. Horses and donkeys seem different, too, but their physical differences are less clear-cut. Mayr thought that species could be defined based on their ability to mate and produce healthy offspring that can also reproduce. Under this definition, which he called the biological species concept, horses and donkeys belong to different species not because they look different but because the offspring of a horse and a donkey is a mule, which can survive but is sterile.

Mayr published his observations in 1942 in a book called *Systematics and the Origin of Species.*[18] It would become one of the foundational texts of modern evolutionary biology.[19] Mayr's detailed knowledge of birds from both his field experience and the extensive museum collections gave him insights into how the subtle variations within any one bird species grade into the more finite distinctions between species. Darwin had suggested that the same process that gives rise to adaptation within a species—natural selection—could also give rise to speciation, the formation of new species. But he struggled to explain exactly how. Mayr found a way to connect the dots.

The key to the formation of new species, according to Mayr, was geographic isolation.

Birds living on different islands that are located close to one another may be able to fly back and forth between islands. If so, then they will keep exchanging genes through sex, and none of these island populations is likely to evolve into a new species. But if the islands are located farther apart, then travel between islands becomes less likely. The evolutionary processes that play out within a species, like natural selection, will happen independently among the populations on each island. Eventually, given enough time, the birds on the isolated islands are likely to become different from those on other islands because they will have accumulated so many differences through processes like mutation and natural selection. Once they have evolved enough difference that they can no longer mate and produce fertile offspring, then speciation has occurred.

In short, Mayr saw that new species come into existence as an outcome of individuals from the same species becoming truly isolated from one another for long periods of time. Islands, by virtue of being isolated, are essentially factories that produce new species.

*

Planets are much like islands. People living in a settlement on another planet, like Mars, would certainly be isolated from people living on Earth. How might living on an isolated planet lead to evolutionary changes in future generations of humans?

For one thing, they might be smaller. We've already considered how natural selection for denser bones to survive childbirth might lead to shorter Martians. But there are also other reasons to expect Martians to change in size.

While collecting data for his PhD on the mammals of a cluster of islands in British Columbia in the early 1960s, biologist Bristol Foster noticed a pattern in the relative size of two species of mice. One was a species of deer mouse that was found on the larger islands as well as on mainland North America. The other was the Sitka mouse, which was found on just a few of the smaller islands as well as on some other islands not far away in Alaska. What caught Foster's attention was the size of the Sitka mouse. It was about twice the weight of the more widespread deer mouse. Based on their similarities, Foster was pretty sure the Sitka mouse had evolved from a common ancestor it shared with the deer mouse, suggesting the larger species had evolved on the islands.

Intrigued about the change in size, Foster began compiling data on the relative body sizes of other island animals around the world. According to his data, island rodents seemed to be consistently larger than their mainland relatives. And the more isolated an island was, the larger its rodents. Other animals, like rabbits and carnivores, showed the opposite pattern—island populations tended to be smaller than those on continents.[20] Foster's paper on the topic focused on mammals, but similar patterns could be seen in other animals. The island of Komodo has its dragons (giant monitor lizards), and Galápagos has its giant tortoises. New Zealand was once home to enormous flightless birds called moas, some of which were twelve feet tall—twice the height of an ostrich! Meanwhile, fossils from some Mediterranean islands showed that they had once been home to pygmy mammoths. Small elephants were also found on islands in Indonesia.

The pattern seemed to be that small animals got bigger on islands while big animals got smaller.

The changes seem connected to differences in the number of predators and the amount of food available on islands compared with continents. The pattern was widespread enough for it to gain traction as a general principle of island evolution, dubbed the "island rule."[21]

*

Could the trends seen in the evolution of island animals also apply to humans? The answer appears to be yes.

People have come to inhabit nearly every piece of land on Earth, including many of even the most remote islands. Indeed, arguably the most remarkable feats of human navigation in history were accomplished by the Polynesian people who settled the majority of the islands in the Pacific. These include the most isolated pieces of land on our planet, like Hawaii and Easter Island. Easter Island, known as Rapa Nui to its Polynesian inhabitants, is only sixty-three square miles in area—a mere speck in the vast Pacific Ocean—and the nearest inhabited island is more than 1,200 miles away. It is the only piece of land within a three-million-square-mile area.

Many people have noted the similarities between the islands of the Pacific and the celestial bodies of outer space that become visible in the night sky. After traveling through Micronesia in the western Pacific, neurologist and author Oliver Sacks wrote that "an ancient practical knowledge . . . had enabled the Micronesians and Polynesians, a thousand years ago, to sail across the immensities of the Pacific by celestial navigation alone, in voyages comparable to interplanetary travel, until, at last, they discovered islands, homes, as rare and far apart as planets in the cosmos."[22]

Intriguingly, the traditional Polynesian worldview makes an explicit connection between islands and the planets and stars in the night sky. Hinano Teavai-Murphy, president of a nonprofit organization dedicated to preserving Polynesia's cultural heritage, explained some aspects of Polynesian cosmology to me during an archaeological tour of the Opunohu Valley on the island of Moorea.[23]

"Each star that is shining in the sky, they correspond to a pillar of the sky," Teavai-Murphy said as we made our way along a steep path through the jungle to the remains of an ancient structure built from black volcanic rock on the green hillside overlooking a picturesque bay.

"We look at the world like it's a bivalve—like a clam," she explained. The top half of the clam shell corresponds to the sky, while the bottom half is the Earth. The pillars hold up the sky. But the pillars were also traditionally believed to be markers of the locations of islands. As the ancient Polynesian voyagers set across the open ocean, they used the locations of stars to guide them to new lands. Assisting them in their navigation, Teavai-Murphy said, was the Polynesian goddess of the Moon, Hina, and her brother, Ru.

"Because she is a woman, [Hina] gives life and she knows where life is," Teavai-Murphy told me. "When she guided the canoes she knows the sea is boiling over there, she knows there is land coming." Hina would tell her brother Ru to pull an island from the sea. "She would hold the island in balance on the surface and the brother would name the island. He is the one who would give the name and by naming the island you give the function of the island," she explained.

Guided by Ru and Hina—and an encyclopedic knowledge of the night sky—the Polynesian people discovered and settled the islands of the Pacific. In many ways, the settling of the remote Pacific Islands was the culmination of the human expansion out of Africa that began some 70,000 years ago. They were the last places on Earth to be settled because they were the most difficult to reach. And yet, despite the formidable distances and the lack of any way of knowing what lay beyond the horizon, somehow people did manage to reach nearly every habitable speck of land across the Pacific. Evidence from archaeology, linguistics, and genetics suggests that the ancestors of the Polynesian people came from Southeast Asia by way of New Guinea. They island-hopped their way across the

Solomon Islands to Samoa and eventually to the Society Islands, which include Moorea and Tahiti, by around 1050 CE. From there, they radiated out to the most distant islands, the last of which were settled by the 1200s.

On the island of Raiatea, in the Society Islands between Moorea and Bora Bora, I visited the site from which many of the long-distance voyages are believed to have departed. The site is called Taputapuātea, and it consists of a series of "marae," platforms made of black, volcanic rock built between the fourteenth and eighteenth centuries.

"A marae is like an open church," our guide explained on the day of my visit. But, she added, it also served as a kind of school. "The marae is also a place where they learn—for navigation, for a future king, for a future warrior, for a future priest."

Taputapuātea is particularly important because it was considered the center of navigational knowledge.[24] The brochure I picked up at the site explained that the island of Raiatea is located "at the centre of the 'Polynesian Triangle,' a vast maritime area dotted with small remote islands between Hawai'i, Easter Island, and New Zealand." Overlaid on an aerial photo of the site was a kind of schematic map, in which the arms of an octopus reached out in all directions toward the various islands—Samoa and Tonga to the west, New Zealand to the south, Easter Island to the east, and Hawaii to the north. At the center was the body of the octopus, labeled with the ancient name by which the island of Raiatea was once known—Havaii.

Voyagers departing on journeys to find new islands would take a stone from the marae at Taputapuātea and use it as the foundation for a new marae wherever they discovered land. The largest marae at Taputapuātea was about 200 feet wide, with a platform made of black volcanic rock and white coral at one end. These represented the land and the sea, respectively, and symbolized the connection between the world of the living and the spirit world. The marae where I was now standing was located on a small peninsula, just a few feet from the edge of the lagoon. The aquamarine water in the lagoon was calm, but looking further I could see the white line that marked the outer reef that surrounds the island. A break in the reef known as Te Ava Mo'a formed the sacred pass, through which voyagers would embark on their journeys.

Looking at the marae and the sea beside it, it suddenly dawned on me—*this was a launch pad.*

In many ways it was much like the launch pads we are building today to send rockets into space, to the Moon, and eventually to Mars. Like many of our modern launch pads, it was located by the edge of the sea. The voyagers who left from this place traveled in double-hulled sailing canoes—their version of spaceships—designed to withstand the perils of the sea. Much like a rocket journey into space, the launch and return were the most dangerous moments of the voyage as the canoes had to pass through the narrow opening between the pounding waves along the reef. And once in the open sea, much like astronauts on a journey to Mars, the voyagers had to be self-sufficient. There was no opportunity to stop for repairs if something broke. They had to bring enough supplies to ensure their survival for a long and unpredictable voyage. Each time the Polynesians set out to settle new lands, they loaded their canoes with everything they would need to start a new home—crops like coconuts, taro, banana, breadfruit, and yams as well as livestock such as chickens and pigs. Martian settlers will no doubt do the same.

*

So, what lessons can we learn from the settling of the Pacific Islands that could inform our future in space?

For one thing, it is clear that the people who settled these islands did not evolve into new species, as many of the animals and plants did. Why not?

One reason is that there was not enough time. The most remote Pacific Islands were settled by around 1200 CE, while the first European explorers arrived in 1520. As European visits to the region became more common in the subsequent years, transportation between islands became easier and more frequent, effectively reducing the extent to which island populations were isolated. Europeans also exchanged genes with Polynesians. That meant there may have been only about twelve generations after the time the islands were settled during which they were truly isolated.

But genetic and archaeological evidence shows that even prior to the arrival of Europeans, Polynesian people regularly traveled between distant islands. One of the most common types of tools used across Polynesia was a stone adze, which was used for many things including felling trees, carving out wooden bowls, and hollowing out canoes. The best stone for this type of tool is fine-grained basalt, which naturally occurs on only a few islands. Yet adzes made with this basalt were found in archaeological excavations on islands across Polynesia, suggesting the existence of widespread trade networks.[25]

Throughout history, when people have exchanged goods, they also exchanged genes.

Indeed, genetic studies of modern Polynesian people support the idea that there was movement of people—and their genes—between widely spaced islands even after they were initially settled.[26] This exchange of genes between populations, which biologists call gene flow, would have reduced the ability of any island population to evolve enough differences for it to undergo speciation.

In other words, as long as people living far apart from one another get together on occasion and make babies, those people will continue to be members of the same species and future generations will continue being able to make more babies together. This is consistent with what Darwin and Mayr concluded from their studies of animals on islands: isolation leads to the formation of new species through natural selection unless gene flow keeps them from diverging.

Genetic studies of Pacific Islanders also revealed another interesting pattern. As the ancestors of modern Polynesians moved east, away from the large and relatively closely spaced islands near New Guinea and into the increasingly scattered islands of the Pacific, they experienced what biologists call a serial bottleneck effect. A bottleneck occurs any time a population is suddenly reduced to a small number of individuals. One way that can happen is if a small group splits off from the larger population, for example, if some members of a group set out into the frontier to settle new lands. If this process happens repeatedly—if the settlers establish a new home and then a subset of them continues on to make yet another settlement, and so on—then you get a series of bottlenecks.

Each time a bottleneck occurs, it has a characteristic effect on the genetic composition of the population. A small group isn't typically representative of the larger group from which it is drawn. By chance, some genetic traits that are present in the larger group won't be present in the individuals who found the new settlement. The genetic diversity of the founding population will be lower than that of the source population, and some rare traits will not be included. Biologists call this the founder effect. When you get new settlements founded one after the other, you get sequential founder effects. Each settlement has less genetic diversity than the population that its founders came from, and one by one the populations become less similar to the original source population.

This is precisely the pattern geneticists observe among Pacific Islanders.[27] Moving from west to east along the path that the ancestors of the Polynesians are thought to have followed, the populations of modern Pacific Islanders have decreasing amounts of genetic diversity. This is because some of the gene variants were left behind as the founders of new populations settled on newly discovered islands.

An example of the founder effect happened more recently in the aftermath of the famous mutiny of the HMS *Bounty*, which took place in 1789. The *Bounty* had been in Tahiti for five months, a prolonged stay that resulted in many of the crew forming relationships with local Tahitian women. After overthrowing Captain William Bligh, a group of nine mutineers and eighteen Tahitians fled to uninhabited Pitcairn Island. Despite a good deal of infighting that included several murders, the surviving mutineers and their Tahitian companions established a small community. Children were born, and the population began to grow. But the island was small, and the limited space and resources soon became a problem. By 1856, the population of 193 people was relocated by the British government to a new island, called Norfolk, where the population again began to grow.

Today, the majority of the roughly 2,000 inhabitants of Norfolk Island can trace their ancestry to the *Bounty* mutineers who escaped to Pitcairn and their Tahitian wives. Genetic studies of the modern population show clear evidence of the founder effect.[28] One study found that among the male inhabitants, 80 percent can trace their ancestry to the one mutineer who was still alive when the population moved to Norfolk. Interestingly, a study published in 2010 found that the reduction in genetic diversity caused by the founder effect has caused the population of Norfolk Island to be somewhat shorter in stature than other populations of Polynesians or Caucasians, the source populations from which the Norfolk population was derived—perhaps an example of the island rule in action.[29]

In addition to changes in genetic diversity, a consequence of the founder effect is that as the number of individuals in the population increases, traits that were once rare can become common and vice versa. This is caused by a phenomenon called "genetic drift" first described in 1932 by American geneticist Sewall Wright.[30] Genetic drift causes traits to become more or less common—not because they are helpful or harmful—but simply due to chance. According to Wright, genetic drift is strongest in small populations—like those recently established by a small group of founders.

The effects of genetic drift can also be seen in populations that stay in one place but are suddenly reduced in size for other reasons. This happened to the unfortunate people living on Pingelap Atoll, in the western Pacific, when a typhoon struck the island in 1775. From a population of around a thousand people, only twenty survived. Among them was the ruler, Doahkaesa Mwanenihsed, who happened to be a carrier for a rare genetic condition called achromatopsia, better known as complete color blindness. The condition is recessive, meaning that to experience symptoms a person must have two copies of the version of the gene that causes it. Mwanenihsed had only one copy and so had normal color vision. But as the population of Pingelap recovered in the years following the typhoon, inbreeding due to the small population size made the rare genetic variant more common. The first people to inherit a copy from both parents, making them colorblind, were in the fourth generation after the typhoon. Today, about one out every ten people on Pingelap is colorblind.[31] By comparison, elsewhere the rate of achromatopsia is about one in 30,000 people.

*

Founder effects and genetic drift can also play a role in the evolution of new species. In a book chapter published in 1954, Ernst Mayr extended his earlier work on the importance of isolation in speciation to include founder effects.[32] He began with an example from his work on the birds of New Guinea. Mayr pointed out that kingfishers are widespread on mainland New Guinea, with three varieties that have subtle differences but are similar enough to be classified as subspecies. Yet the smaller islands surrounding New Guinea contain species of kingfishers that are strikingly different from one another and different from those on the mainland.

For instance, the little island of Numfor is home to a species of kingfisher with a blue breast, very distinct from the white-breasted kingfishers on the mainland. Mayr assumed the ancestors of the Numfor kingfishers came from New Guinea and initially had very limited genetic diversity. The blue color of the island population's breast feathers wasn't necessarily because blue feathers were advantageous on Numfor; rather, there happened to be some genes for blue breast feathers among the founding population, and those genes became more common due to chance as the population grew and genetic drift kicked in. Eventually, all of the kingfishers on Numfor had blue breasts, and perhaps other traits. According to Mayr's definition, the process of speciation was complete once the individuals no longer mate and produce fertile offspring, which seemed to be the case for the Numfor and New Guinea kingfishers.

Subsequent research has shown that Mayr's concept of founder speciation is not the primary way that new species evolve, as Mayr originally proposed in his 1954 paper, but it does happen.[33] Not surprisingly, it is more common among species living on islands.

For example, there are about 1,600 species of fruit flies in the genus *Drosophila* living around the world, but about a quarter of them live in the Hawaiian Islands. The reason there are so many fruit fly species in Hawaii is that the Hawaiian fruit flies went through a series of founder speciation events. New species came into existence as each of the volcanic islands emerged from the sea and was colonized by a small number of fruit flies, each of which led to an instance of founder speciation.[34]

Galápagos tortoises also evolved through a series of founder speciation events. Much like the Hawaiian Islands, the Galápagos Islands were formed by a volcanic hotspot. The first of the tortoises that arrived in the Galápagos, presumably by floating over from mainland South America, arrived on the islands that are now the oldest in the archipelago around 3 million years ago.[35] The founding individuals were much smaller than any of the modern Galápagos tortoises—their closest living relative, the Chaco tortoise, weighs only about three pounds. Following the island rule, once isolated the tortoises evolved to become the giants we know today, which can weigh up to 600 pounds.

As new islands formed in the Galápagos chain, a small number of tortoises—perhaps as few as just one gravid female—found their way to them. In each case, the founding individuals established a population that evolved into new species, leading to a total of sixteen species. Among the differences between them are shapes of their shells—a fact that proved useful to Darwin.

✷

People living on islands like those in the Pacific were never isolated long enough to evolve into new species following the patterns seen in tortoises, fruit flies, birds, and other animals. But that doesn't mean that humans are not subject to the same evolutionary forces as other species.

Evidence for this was discovered in 2003, when researchers uncovered a remarkable human-like skull while excavating in a cave on the Indonesian island of Flores.[36] The skull was small, leading the researchers to assume at first that it was a child. But as the excavations continued it became clear that the skull and mandible included the teeth of a typical adult. Bones from the lower body were discovered as well, revealing that these were the remains of a previously unknown species of human who stood only about three and a half feet tall. Additional fossils, including teeth and an arm bone found at another site nearby that date to 700,000 years ago, were reported in 2024.[37] The arm bone suggested an even shorter stature for the species. Its scientific name, *Homo floresiensis*, was based on the island where it was discovered. But the unofficial nickname its discoverers used, "hobbits," captured the intriguing uniqueness of the species, which lived between 190,000 and 50,000 years ago.

The fact that the fossils were on Flores was particularly interesting. While some of the islands in western Indonesia, like Java and Sumatra, were once connected by land bridges to mainland Asia, Flores and its surrounding islands remained separated from Asia by stretches of deep water. That meant that the ancestors of *Homo floresiensis*, possibly *Homo erectus*, must have somehow crossed the open ocean to reach Flores. Whether they did so intentionally, perhaps by building some sort of raft, or whether they were accidentally swept out to sea, is unknown. However they arrived, once there they would have been truly isolated.

Intriguingly, the fossilized remains of another small-bodied human were discovered in 2019 on the island of Luzon in the Philippines. Named *Homo luzonensis*, the remains were dated to around 67,000 years ago, although stone tools from the same island are 700,000 years old.[38] Luzon was never connected to mainland Asia, nor was it connected by land bridge to Flores. It seems possible that the humans on both Flores and Luzon evolved into new species as a consequence of their isolation—perhaps for several hundred thousand years.

No DNA has yet been retrieved from any of the ancient human remains from these islands, so we do not know for sure how they were related to each other or to other human species such as *Homo erectus*, or how much genetic diversity either species contained. But we can speculate based on how difficult it must have been to get to both Flores and Luzon that the founders of each population would have been a small number of individuals. They would have been subject to the founder effect, reducing the genetic diversity of the population. Later, genetic drift would have randomly caused some traits to become more common and others to be rare. Their small body sizes match the predictions of the island rule.

Humans, it seems, become smaller when isolated.

*

Would Martians become smaller, too?

They might. One reason large animals are thought to become smaller on islands is that they often have restricted access to food. That could very well be the case for humans living in space settlements, as we will see in the next chapter. Natural selection might favor individuals who require fewer calories if they have a survival advantage over their hungrier neighbors, potentially leading to smaller body size and lower metabolic needs in their children.

One lesson from island studies is clear: founder effects and genetic drift are likely to play major roles in any human space settlements. Given this, the exact size of the founding population will be a critical factor that should be given careful consideration. In general, the larger the founding population, the better. Observations of human groups living traditional hunter-gatherer lifestyles have prompted some to suggest that a population of 150 individuals is sustainable. While a founding population this size might survive, it would experience a strong founder effect. And as the population expanded, it would be subjected to high levels of genetic drift, making it more likely for some traits initially present among the founders to be lost. Meanwhile, other traits—including potentially harmful ones—could become fixed, meaning that they are present in every individual in the population. A study by space anthropologist Cameron Smith concluded that a safe number for a founding population of people to survive for at least five generations, accounting for the possibility of a few population crashes, would be between 10,000 and 11,000 individuals.[39]

In addition to the size of the initial population, the likelihood of founder effects means that diversity will also be an essential consideration. All aspects of human diversity will be important, so including people who differ in their culture, language, ethnicity, skills, physical abilities, artistic talent, and so on will aid in the success and vibrancy of a space settlement. But genetic diversity will be particularly important since it will affect the potential for future evolution of the population. A population of about 10,000 founders is the minimum size that could conceivably capture all of the genetic diversity among people alive today on Earth, if they were selected with genetic diversity in mind.

Put simply, maximizing genetic diversity will maximize the chances of long-term success for future generations of Martian settlers.

This is partly because genetic diversity is the raw material that natural selection uses for adaptation. To use a silly example, no matter how useful it might be for a person to have wings so they could fly, unless there are at least some individuals with genes that give them wings, nobody in later generations will be able to fly. Natural selection can only use what already exists in the population and make certain traits more or less common.

If the goal were to maximize the genetic diversity of the founders of a settlement on Mars, this would lead to some interesting changes in the selection criteria compared with how astronauts have historically been chosen. In the early days of the American space program, the people who became astronauts were chosen from an already elite group of military test pilots. Notably, they were all white males. It took a long time for NASA to diversify its astronaut corps, but even today those who are selected are very far from being a truly representative slice of the American public. Of course, that has never been the goal. NASA wants astronauts who are the most likely to be successful in their jobs. Since they'll be subjected to very physically, psychologically, and socially stressful situations, people are chosen who have proven themselves to do the best in those circumstances. By doing so they tend to get people with a narrow range of certain types of human traits and abilities.

The founders of a space settlement would need to be highly competent, too, of course. In many ways the physical, psychological, and social stresses of starting a settlement on another planet will likely be far more extreme than what any previous astronauts have experienced. Yet if the goal is not only to survive but also to lay the foundation for future generations of Martians, then we will need to cast a much wider net when selecting candidates.

Ideally, the founders should not be drawn from any one nation but should represent people from a great many countries around the world. Human genetic variation is not distributed evenly across the planet, so even just choosing people from many nations would not be sufficient. For example, it would be possible to choose people from the nations that have historically dominated space exploration, such as the United States, Russia, Japan, and various European countries, and still have a very small and unrepresentative slice of human genetic diversity. By far the greatest amount of genetic diversity is found in Africa—in fact there is more human genetic diversity in Africa than in the rest of the world combined.[40] This is due largely to the fact that Africa is where our species has had the longest history. As such, it could be argued that the best and simplest way to select a group of candidates for a Martian settlement would be to have them all be African. At the very least, a good number of them should be African or members of the African diaspora.

Inevitably, discussions of who should be selected as the founders of a new human population will lead to ethical considerations. For example, does maximizing genetic diversity mean including all aspects of human genetic diversity, including individuals with genetic conditions like cystic fibrosis or sickle cell disease? What about people with genetically based hearing loss or infertility? At the heart of these considerations are questions about what aspects of human variation we consider valuable and important. And who gets to make these choices? We'll return to these questions later.

In addition to considerations about how to select the initial founders of a Martian settlement, studies of life on islands make it clear that gene flow between isolated populations is a major factor in whether an island population evolves into a new species. The take-home message from all our island exploration is this: the more movement of people back and forth between Earth and Mars, the less likely it will be for people on Mars to evolve into a new species. That is, as long as people from both planets exchange genes by having children together, a new species of human is less likely to evolve on Mars. To undergo speciation and truly become Martians, something would need to limit reproduction between Earthlings and Martians.

That something might come from the things we bring with us, whether we realize it or not.

7

OUR ENTOURAGE

Human settlements on Mars will be different in many ways from the space stations we have been inhabiting for about three decades and even from the bases currently being planned on the Moon. One of the main differences is the distance. The International Space Station is located about 250 miles from Earth and can be reached in a matter of hours. The Moon is about 1,000 times farther away but can still be reached in about four days. By contrast, the closest Mars gets to Earth is about 35 million miles, and its orbit takes it as far as 250 million miles away.[1] When the planets are at their closest, which happens about every two years, it takes about seven to eight months to get from Earth to Mars using current rocket technology. This means that sending regular shipments to a Martian settlement containing supplies—including food—will not be practical.

Rather, people living on Mars will have to be self-sufficient. They will need to use the resources available on the planet and will have to produce their own building materials, medicine, and food. The fact that Mars has natural resources like water, carbon, and oxygen as well as materials that can be used for building and manufacturing, like iron, titanium, aluminum, and silicon are what make Mars the most attractive place we know of in our solar system for human habitation.[2] On the other hand, as far as we know Mars is currently a lifeless planet. That means that in order for us to live there, we would need to build an entire functional ecosystem essentially from scratch.

This is more difficult than many proponents of space settlement seem to realize. One of the most avid advocates for settling Mars, Robert Zubrin, wrote,[3] "Mars appears barren to most people today, just as Ice Age Europe and Asia must have appeared to early humans migrating out of our original tropical African natural habitat. Yet, by developing new technologies, new attitudes, and new customs, our ancestors were able to create the resources to not only sustain themselves, but also to flourish with ever-increasing prosperity across the entire planet."

But here's the thing: even Ice Age Europe and Asia had basic resources that people could use. As the glaciers retreated, they left flowing water that people could drink and which attracted animals that could be hunted and killed. Plants grew as the soil was replenished, providing other sources of food. Fish and other aquatic animals could be harvested along coastlines. And although they couldn't see it, the soil, water, and air were all teeming with beneficial microorganisms. Not to mention that there was plenty of oxygen to breathe. The first settlers on Mars will have none of these benefits as they embark on creating a livable environment.

Let's focus on our food supply. All of our food comes from living things, mostly plants and animals. The United Nations estimates that 61 percent of the calories consumed worldwide come from crop plants, while another 14 percent comes from livestock like cattle, pigs, and chicken.[4] Much of the rest comes from wildlife, especially fish. There is a compelling case to be made that people living in space settlements will have to be vegan. The reason is simple. If you eat meat, you are getting your calories from an animal that was likely consuming a largely plant-based diet. If you are vegetarian—but not vegan—then fewer of your calories come from animals, but products like eggs or dairy come from animals, which again get their calories from plants. Vegans get all of their calories directly from plants. Regardless of your dietary choices, all humans on earth consume calories from plants, either directly or indirectly. But only 10 percent of the energy from plants makes it into a person or animal that eats those plants; the rest is used by the plants for their own metabolic needs or is lost to the environment as heat. If we eat corn, we get 10 percent of the energy that the corn plant acquired from sunlight through photosynthesis. But if we instead feed the corn to cattle, and then make a hamburger from the cattle, we are only getting 10 percent of the 10 percent—just 1 percent—of the energy from the corn. That kind of energy inefficiency might not seem like a big deal on Earth, but in a Martian settlement, where space and energy will be at a premium, such waste will not be an option.

There are other reasons why bringing animals to Mars as livestock won't be a good idea. Their size and weight are one. It simply isn't cost effective to put anything as large as a cow—even a young one—on a rocket headed to Mars. Besides the cost of getting it there, any animals we bring would compete with us for valuable resources like oxygen, water, food, and space. Livestock like cattle, pigs, and goats need a lot of food and water, and they consume lots of oxygen, too. Even smaller livestock, like chickens, would still compete with us for resources that are keeping us alive. As much as it pains me to say this as a meat eater, we would be better off without them.

Although, if animal protein were considered a necessity, then one option would be insects. While people from North America and Europe might consider it odd or objectionable to eat insects, they're a part of the diet for more than two billion people around the world. Insects are nutritious—mealworms and crickets, which are farmed worldwide, are comparable in protein content to beef but have less fat. Unlike large livestock like cattle and pigs, insects can also be raised in small, indoor enclosures. Insects also consume substantially less water and food—producing one kilogram of beef requires ten kilograms of feed, while the same amount of crickets can be produced using just 1.7 kilograms of feed.[5]

*

A plant-based diet, perhaps supplemented by insects, is the logical choice for space settlements. But can we grow plants on Mars?

Until we get our terraforming project far enough along that we can grow plants outside, all the growing will have to be done indoors, in greenhouses. That will likely require artificial lighting, although some natural sunlight might be usable. But as any gardener knows, the real key to success when growing plants is having good soil. Technically there isn't *any* soil on Mars, since soil is a combination of minerals and organic matter from the decomposed remains of plants and other organisms, as well as lots of microorganisms. The surface of Mars is devoid of both organic matter and microbes; that's why researchers prefer the term "regolith."

Okay, so can plants grow on Martian regolith? We don't know for sure because we have never been able to try. But a fair amount of research has been done using *simulated* Martian regolith—rock from somewhere on Earth that has similar physical and chemical properties to what's on Mars. Results of studies growing plants in simulated Martian regolith have been somewhat mixed. A study by Dutch researchers published in 2014 found that several types of plants could germinate and grow in simulated Martian regolith from Hawaii. Among the crop species in their experiment, rye and tomatoes grew particularly well, but after fifty days none of the tomatoes produced flowers. The rye and two other types of plants produced flowers, but none of them made seeds, which would be problematic for keeping a Martian farm going.[6]

None of the simulated Martian regolith contains perchlorates, which our rover studies show are widespread on Mars. Plant growth experiments in which perchlorates were added to simulated Martian regolith showed that seeds can germinate but do not survive long after that.[7] But even if perchlorates can be removed, for example, by rinsing the regolith with water, the regolith would need to be supplemented with some form of fertilizer. Anyone who has read Andy Weir's novel *The Martian* or seen the movie version starring Matt Damon is familiar with the idea that human waste could be used as fertilizer for Martian crops.[8] Experiments using simulated Martian regolith—and even in actual lunar regolith from the Apollo missions—found that plants grow only when given fertilizer. I am disappointed to report that, as far as I can tell—and I looked—none of the published experiments growing plants in simulated Martian regolith have used human waste as fertilizer.

However, the Dutch researchers performed a follow-up study to their 2014 paper in which they grew ten types of crop plants using a nutrient solution "to mimic the addition of human faeces and urine." They also added chopped-up bits of rye grass to the regolith to boost its organic content in a way they imagined would be similar to how Martian farmers would enrich their soils with compost from previous harvests. With this approach, they were able to grow and harvest edible parts from six different types of crop plants, and found that plants grew as well on the simulated Martian regolith as in Earth soil.[9]

Other studies have had similarly encouraging results when using fertilizer with simulated Martian regolith.[10] However, one aspect of these studies that is worth considering is that the simulated Martian regolith was not sterilized before the experiments began. That means that they likely contained microorganisms, which can be helpful for plant growth. Here on Earth, it's easy to overlook the importance of microbes—not only because they are too small for us to see but also because they are just so ubiquitous. A small scoop of topsoil from a typical garden has upward of 5 billion fungi, bacteria, and other microbes.

Indeed, many plants require microorganisms in order to survive. These include fungi known as mycorrhizae, which form symbiotic relationships with plants in which the fungi grow into the plants' root tissues and help the plants to better absorb water and nutrients.[11] In return, the plants supply the fungi with energy in the form of sugar. Mycorrhizal fungi are especially important for plants growing in nutrient-poor soils. Crop plants are often inoculated with mycorrhizal fungi to boost their growth. However, the fungi naturally occur in most soils on Earth, so plants can often acquire them on their own and farmers only need to inoculate their soil when the mycorrhizae have been depleted by tilling or other processes.

There are no mycorrhizal fungi on Mars, so we will need to bring them with us to help our crop plants grow. Other types of microbes will also need to be imported to Mars in order to help us grow food there. Some fungi and bacteria live as symbiotic partners in the above-ground parts of plants like leaves. These are called endophytes and they also play important roles in helping plants to get nutrients and manage stress, like during droughts.[12]

A particularly helpful group of bacteria, the rhizobia, live in nodules on the roots of legumes, like beans and peas. Legumes are generally high in protein so they will likely be a key component of the Martian diet. The rhizobia are important because they convert nitrogen in the air into a form that can be used by plants. Because of the activity of these bacteria, planting legumes increases the amount of nitrogen in the soil, which is the reason that legumes are often planted together with other types of crop plants or in the same soil during alternating seasons. Having rhizobia reduces the amount of fertilizer needed to keep crops healthy—but again, we will have to bring them with us if we want them on Mars. With very little nitrogen in the Martian atmosphere, we will want our plants to take advantage of as much of it as they can get. Encouragingly, one study published in 2021 found that legumes and their rhizobia could grow in simulated Martian regolith.[13]

In short, to grow food on Mars we will need to bring a veritable microbial ark.

✷

But we have a long way to go before we can confidently load such an ark. The reason is we don't yet know the full set of microbes our crops plants will need to survive on Mars. And we are only beginning to look into whether any of those microbes will be able to survive there.

But the bacteria and fungi our plants need won't be the only microbes that we bring with us. When we travel to Mars, we will carry microbes in and on our bodies. The realization that our bodies harbor an entire microbial ecosystem, called the human microbiome, has led to something of a revolution in the fields of biology and medicine in the last few decades. The development of so-called next-generation DNA sequencing technology in the early 2000s made it possible to identify microorganisms based only on their genetic material. Now, by taking a small sample—a swab from a cheek or an armpit, say—researchers could identify the various microbial inhabitants of that part of the body just by sequencing the sample's DNA.

This was a major improvement over the classical methods of microbiology, which involved attempting to grow any microbes from a sample in a laboratory culture, like a Petri dish. It turned out that many types of microbes won't grow in laboratory cultures, or at least we don't yet know what they need to stay alive and grow. That meant that many earlier studies of microbial diversity, including that of the human body, were overlooking vast numbers of microbial species. We simply had no idea they were there.

Microbiome studies using the new approach soon revealed that the single-celled microbes in and on our bodies are so numerous that they outnumber our own cells. Their identities vary, both from person to person but also from one body part to another. And the precise breakdown of how many of this type versus that type also fluctuates within an individual person depending on things such as what we eat, where we travel, and how often we use antibiotics. Our microbiome also changes as we age.

Figuring out not just the identity but the *functions* of all these microbes has been even harder, but there have been some major revelations. First, the vast majority of the microbes in and on our body are not making us sick. We used to think of all microbes as germs, but that is an outdated perspective. Most of the microbes that live with us are either harmless or helpful. Those that are helpful are perhaps the most interesting. They do things like help us digest food, regulate our moods, and fight off actual germs—the microbial pathogens that cause disease. We depend on these helpful microbes to perform many essential tasks. In effect, we have outsourced some of our physiological functions to the hardworking microbes that inhabit our bodies.

Because we bring our microbiomes with us everywhere we go, there is no doubt that they come with us even when we leave Earth. As science writer Ed Yong put it,[14] "When Neil Armstrong and Buzz Aldrin set foot on the Moon, they were also taking giant steps for microbe-kind."

Indeed, studies of astronauts have shown that the human microbiome does indeed follow us into space, but our microbiome is also affected by spaceflight. In the NASA Twins Study, stool samples were collected from Scott Kelly during his year on the International Space Station, as well as from his brother Mark Kelly back on Earth. Interestingly, the composition of Scott's and Mark's gut microbiomes were somewhat different at the start of the study. This wasn't too surprising, because other studies of twins have found them to have similar, yet distinct microbiomes. Despite being genetically identical and growing up in the same home, any slight differences in their food preferences, antibiotic use, and other life experiences can contribute to differences in the precise makeup of their gut microbial community.

As expected, the composition of Scott Kelly's gut microbiome changed during his year in space more than did his brother's during the same time period.[15] In general, a diverse gut microbiome is considered healthy. While the overall diversity of microbes in Scott's gut did not decline, certain types of bacteria became more or less common during his time in space. A common metric used to examine gut microbiome health is the ratio of bacteria in the group known as the Firmicutes compared with the group known as the Bacteroidetes. While the total number of each type of bacteria can change over time, the ratio usually stays about the same in healthy adults. In Scott Kelly's gut, the ratio went up while he was in space, meaning there were relatively more Firmicutes compared with Bacteroidetes. This imbalance can be associated with a wide range of diseases, from irritable bowel syndrome to rheumatoid arthritis, along with many others. Fortunately for Scott Kelly, he did not experience any of these conditions.

Some of the differences in Scott Kelly's gut microbiome could be related to what he ate in space. After all, fresh food—which often comes with a healthy helping of microbes—is a very rare luxury on the International Space Station. Most astronaut food is pre-packaged and treated with radiation to help preserve it, minimizing the number of lingering microbes that could make their way into the gut. Another possible factor is the fact that Scott was confined to a small living space with just a few other people for the duration of his mission. The changes to his gut microbiome that occurred in space were reversed just a few weeks after he returned to Earth.

While the Twins Study focused on the gut microbiome, a study of how other parts of the human microbiome are affected by spaceflight was done during the Inspiration4 mission in 2021.[16] The crew members took swabs of their mouth, nose, and skin from eight different parts of the body before, during, and after the flight. There were changes in the microbiomes of all of these locations while in space. Some types of bacteria and viruses became more common, while others became less common. It was interesting that even a few days in space resulted in notable changes in the microbiomes of the four civilian astronauts. However, as was seen in the gut microbiome in the NASA Twins Study, many of the microbiome changes observed in the Inspiration4 crew were temporary and went back to baseline after they returned to Earth.

The microbiomes of the Inspiration4 crew members, especially that of the skin, became more similar to one another during the flight. This suggests that the crew members were exchanging microbes, most likely because they were all living together in the same small SpaceX Dragon capsule. Swabs of the capsule itself showed that the crew members were also exchanging microbes with the surfaces they were touching. The capsule was swabbed twice during the flight, and the microbes living on it were more similar to the microbes on the crew members during the second swab, showing that the more time they spent in the capsule, the more they swapped microbes with the spacecraft.

NASA has conducted microbiology studies since the early days of the space program. During the Apollo missions, the primary concern was preventing the possibility of microbes from the Moon being brought to Earth. The Apollo astronauts who landed on the Moon spent three weeks in quarantine after their return.[17] On Skylab there were more detailed studies of the microbes that astronauts bring with them to space, including surveys of the spacecraft.

While these studies relied on traditional laboratory culture approaches to microbiology, and therefore would not have been able to detect all of the microorganisms that were present, they could at least make meaningful comparisons. For example, the amount of bacteria sampled from the air within Skylab increased over the span of the three missions. In their official report, NASA researchers attributed this increase to the arrival of one species of bacteria, *Serratia marcescens*, which can cause a variety of human diseases, including respiratory infections and meningitis.[18] The authors noted that it had not been seen in Skylab or any of the Skylab crew prior to the final mission, nor was it detected in any of the crew prior to their flight. However, they wrote that "this species persisted in the nasal cavity of the Pilot throughout the postflight quarantine period." Samples collected from the surfaces within Skylab showed similar increases in the amount of bacteria during the missions. Importantly, the authors noted that the spacecraft that was used for the Skylab Orbital workshop was not devoid of bacteria prior to launch, but had levels of bacteria they described as "typical of a clean environment."

Microbiology studies were also conducted on the Russian space station, Mir. It was not what anyone would describe as a clean environment. Mir was overrun with fungi—so many that it was said to smell like rotting apples. Astronaut David Wolf described helping to clean up a floating ball of "slimy goo" during his time on Mir, which was apparently not unusual. Trying to make a good impression on his Russian companions, Wolf offered to take on the task of cleaning up these floating blobs anytime they found any. That turned out to be more often than he expected.

"I didn't realize what I was getting myself into, because it took anywhere from four to eight hours a day, the rest of the mission, every single day except a few," he recalled.[19] The blobs got big, too: "bowling ball or beach ball size sometimes," Wolf noted. He brought back some samples of these blobs, which turned out to be chock-full of microbial life.[20] There were at least seven different kinds of fungi, dozens of bacteria, and even some amoebae and dust mites. No wonder it smelled rotten.

✷

Studies onboard the International Space Station, particularly those in the era of next-generation DNA sequencing technology, have given more insights into the microbes that inadvertently get carried into space.[21] To get a sense of what we've learned from these studies, I made a visit to Johnson Space Center to meet Sarah Wallace, who runs the microbiology lab.

When I stepped into her office, the first thing I noticed was all the magnets. "Travel is my passion," Wallace told me. Everywhere she goes, she brings back a magnet or two. There were magnets from the Badlands of South Dakota, Scotland, and Bora Bora. There were also some with more mysterious origins. "I love Bigfoot, Jurassic Park, dinosaurs, Harry Potter . . . all the nerd things," she said with a laugh, pointing to a few of her favorite magnets.

I asked her where her interests came from. The Bigfoot thing came from her grandparents. "My grandma was a big believer," she said. Wallace likes to keep that open-minded perspective, even as a serious scientist. "Sometimes you get so rooted in the science . . . especially when you know your area really well . . . you start to limit yourself," she said. "But with Bigfoot, and the Loch Ness monster, who knows, right?"

Wallace grew up in the small town of Goddard, Kansas. Her sixth-grade science teacher got her excited about space. "We were quizzed over NASA history," she recalled. "Some kids probably thought it was boring and awful but I'm like, oh, this is the greatest thing I've ever heard!" she said. As a teenager, she went to Space Camp, which included a trip to Johnson Space Center. She decided at that moment that she wanted to work for NASA.

We left her office and she led me through a door into a laboratory. There were black lab benches with incubators, racks with pipettors, and shelves filled with stacks of Petri dishes. "Our job is to keep the vehicle clean and keep our crew healthy," Wallace explained as we walked through the lab. "For my risk assessment, when I say thumbs up or thumbs down, I want two things—and that's 'about how much,' and 'what is it'?"

She explained how the crew on the space station follows standard procedures to take samples of the air and water and swab various surfaces. Then they place those samples into a device that allows microbes to grow so they can be identified. She showed me an example of what one of the devices used for surface samples looks like. Unlike the standard round Petri dishes used in laboratories on the ground, where size is not a factor, the culture dishes on the space station have to be tiny. One of them looked like a pack of chewing gum, with individual pale-yellow rectangles, each of which is a miniature culture dish. After five days, the crew look at each of the little culture rectangles and call down to Mission Control to let them know what has grown. Wallace showed me the scale diagram they use, in which the density of bacterial growth ranges from a 1 (very little) to a 4 (almost completely covered in bacteria).

"They simply look at what grows and they say, 'it looks like a one,'" she explained. The same thing is done for fungi, except with an alphabetical scale to avoid confusion. Sometimes Wallace's team will ask the crew to take photos of the plates to confirm the astronauts' assessments. While they can't identify the microbes until they get them back to Earth, they can at least monitor for any changes. "It gives us a good idea of, okay, this site was really dirty—you guys need to up your housekeeping game," she said.

While Wallace is pushing to be able to do more on board without having to bring samples back, she is proud of what their protocols have been able to achieve. "This has served us really well. We've kept a healthy crew. We've kept a healthy vehicle . . . from a microbial standpoint, things are great," she said.

Wallace showed me a machine that can automatically identify species of bacteria based on their biochemical properties, and was developed as a spin-off of NASA technology. But it's the next-generation DNA sequencing technology that she finds especially exciting. We stepped through another door and into the molecular lab where the sequencer machines are kept. "To me, this is better than Building 9, which is the building with all the [space vehicle] mockups," she said with a laugh.

She showed me three types of progressively smaller DNA sequencing machines, each of which uses a different technique. The last one, called a MinION, looked small enough to fit in my pocket. It was used to perform the first-ever DNA sequencing experiment in space, conducted on the International Space Station by astronaut Kate Rubins in 2016. It worked perfectly—a little better, even, than on the ground. Over the next eight years an additional twelve astronauts have used the methods developed by Wallace and her team onboard the International Space Station to identify microbes both with and without growing them first. Having the ability to sequence the DNA of microbes without having to culture them first was a game changer and is how NASA plans to continue monitoring microbes in the environment for the missions being planned on the Moon and, eventually, on Mars.

✷

One of the first big studies of microbes on the International Space Station in this new era was led by microbiologist Jonathan Eisen and published in 2017.[22] As expected, Eisen and his collaborators found types of microbes that had not been detected in previous research using traditional techniques. An interesting aspect of this study was that the protocol followed as closely as possible one that had been developed by biologist Rob Dunn to study the microbes in people's homes. Dunn described the results in his book *Never Home Alone.*[23] "In nearly every way, the bacteria of the [International Space Station] are the sorts of bacteria we would expect in a house on Earth if all the environmental influences were removed," Dunn wrote. "[The International Space Station] is what you get when you scrub and scrub and close the windows, doors, and hatches."

Many of the same microbes found in people's homes were also on the International Space Station, even though there were fewer microbes overall. The closest matches were bacteria commonly found on pillows and toilet seats, which Dunn points out "are different from one another, but not as different as you might hope." Indeed, many are bacteria associated with human feces. But pillows also tend to have bacteria found in saliva (yes, you drool when you sleep) but those were less common on the space station. All of the fungi found on the space station were also in people's homes. On the other hand, some of the most common bacteria in the space station are the types commonly found on human skin. Taken altogether as a community, the microbes on the International Space Station turned out to be more similar to the microbial communities in people's homes than they were to the communities that make up the human microbiome.

Similar results were found in a study published in 2019, which used both next-generation sequencing and traditional culture-based microbiology techniques to examine the microbes living on surfaces within the International Space Station.[24] The majority of the microbes they detected—64 percent of the bacteria and 60 percent of the fungi—could not be cultured and therefore would have been missed if they had used only traditional techniques. They also found that the communities of microbes on the International Space Station were similar to those found on surfaces in homes, offices, and labs back on Earth. Similar, but not identical. The microbial communities growing inside the International Space Station are in fact unique.

That should make you pause and think.

Many of the microbes in people's homes come from the people who live inside them. But microbes also come into our homes from the outside, whether through doors and windows or on the soles of our shoes when we step inside. Our homes are surrounded by a microbial jungle. The International Space Station is not. The only possible ways that microorganisms can get to the space station are by hitching a ride on people or on cargo. All cargo undergoes a thorough decontamination process before it can be sent to the International Space Station, so there should not be many microbes arriving that way. You can't do that to people, so they must be the ones shepherding microbes up to space. But then the microbes on the space station should look like the sort that are typically found on our bodies. How, then, could the microbial communities found on the space station appear to be so different?

Here's the thing. It's not necessarily the case that the species on the International Space Station are different from the species in people's homes or on their bodies. What the studies found is that the relative proportions of species differ between all three, but that the proportions are most similar between homes and the space station. What these studies couldn't do is examine all the individual species to see if they were exactly the same or not. Rather, in most cases they identified each microbe to its genus, the broader category to which biologists classify species. That is usually good enough for most studies comparing the microbes in different places.

But there is another possibility. What if the microbes on the space station have changed during their time in space? Perhaps the microbes living on the International Space Station today are not exactly the same as those that first came aboard a couple decades ago with the first few crews. In fact, the 2019 study collected samples from the International Space Station at three different times over a span of fourteen months. They found that the bacterial communities changed over time, with certain types becoming more common, while others became less common.

But it is also possible that the microbes themselves are changing over time—that is, that they are evolving.

In fact, we know they are. One species of bacteria, *Acinetobacter pittii*, that has been found growing on surfaces within the International Space Station has shown clear signs of evolutionary changes. A study led by researchers at Cornell Medical School and NASA's Jet Propulsion Laboratory found that the samples of *Acinetobacter pittii* from the space station had genetic changes that have not been found in any samples of the same species on Earth.[25] When the researchers constructed an evolutionary tree that shows how the individual bacteria are related to one another, they found that the *Acinetobacter pittii* individuals on the space station are all closely related to each other and that they evolved from a common ancestor—presumably a small number of *Acinetobacter pittii* bacteria that were inadvertently carried to the International Space Station on one of the astronauts' bodies.

Some of the changes in the *Acinetobacter pittii* from the space station seem to make it better adapted for living in space—just as Darwin would have predicted. For example, there were changes in genes that are involved in repairing DNA damage, like the kind caused by radiation. There were also genetic changes that made the bacteria more resistant to certain antibiotics—a disconcerting finding given that the Earth-bound version of *Acinetobacter pittii* is known to cause deadly infections in people with weakened immune systems.

Another new species of bacteria discovered on the International Space Station, *Methylobacterium ajmalii*, has so far not been found anywhere on Earth.[26] This raises the intriguing possibility that it could have originated on the space station. In other words, the ancestors of this species may have been another species of *Methylobacterium* that was transported to the International Space Station and went through the process of speciation while on board. Supporting this possibility is the fact that the newly discovered bacteria had evolved changes to the genes that improve their ability to tolerate stressful conditions, which the bacteria no doubt have faced since being in space.

Other types of bacteria have also been found evolving on the International Space Station, even if they have not (yet) evolved into new species. Like *Acinetobacter pittii*, several have developed antibiotic resistance. Even more concerning is a trend in which some microbes in space have evolved to be more virulent, meaning that when they cause an infection, they make their host sicker.[27] Part of the reason for this appears to be that some microbes grow better in microgravity, which allows them to spread more rapidly. Fortunately, this has not resulted in any dangerous infections for people on the International Space Station, at least not yet.

But infections are a big concern for people during spaceflight—not only because the microbes that cause infectious diseases seem to get nastier in space, but also because the human immune system seems to be affected by spaceflight. During the Apollo and Skylab missions, most of the data on astronauts' immune systems came from studies that were done after the astronauts returned to Earth. They showed clear signs of having altered immune system function, and often came down with mostly minor illnesses once back home, but it was hard to tell whether that was caused by the stressful reentry process. In later studies, blood tests taken while astronauts were aboard the space shuttle showed that, while there were some impacts caused by reentry, various aspects of the astronauts' immune systems were out of whack while they were in space. Specifically, the levels of certain components of the immune system, like T cells and cytokines, were found at higher or lower levels than before or after the flight.[28] These initial studies were confirmed and expanded upon by more detailed analyses on the International Space Station, including the Twins Study. Studies of the Inspiration4 crew showed that even a three-day spaceflight activates some of the same genes observed in the Twins Study, including genes that play important roles in the immune response.[29]

One conclusion from these studies was that certain aspects of the immune system, like white blood cells that respond to an infection, do not always work as well in space. The effect appears to be caused at least in part by reduced gravity. This can make people more vulnerable to infection by microbes they come into contact with. It can also make them more vulnerable to those already in their bodies. One common outcome for astronauts during spaceflight has been an awakening of viruses that were dormant within their bodies.[30] Most of us carry these so-called latent viruses around even when we are healthy, particularly herpes viruses like the kind that causes chickenpox and shingles. We don't notice them because they don't have any effect on us as long as we are generally healthy. But they can become reactivated if our immune systems are weakened, as often happens in space. Luckily for astronauts, the consequences of reactivating latent viruses have so far been minor, just some skin rashes and minor allergies.

But these viruses and their effects on astronauts should not be our main concern—rather, they are the canaries in the coal mine. The fact that spaceflight impairs the immune system, combined with the observations that microbes are evolving to be more virulent in space, should be a red flag. Or at least it should make us think carefully about our plans for the Red Planet.

*

It is clear that microbes come with us when we go to space. The microbial entourage we call our microbiome will travel with us when we go to Mars. We will need to carefully select those microbes we want to bring to help our plants to grow. A lot of other microbes will come along as stowaways, as they have already done on our spacecraft. Altogether, the microbes we choose as well as any hitchhikers will plant the seeds of a new ecosystem.

But the ecosystem we create on Mars will contain only a small fraction of the microbes we are normally surrounded by.

On the one hand, this will be an enormous opportunity. If we are cautious about screening people traveling from Earth to Mars for any infectious diseases, it should be possible to prevent many of those diseases from getting a foothold in our new settlements. Quarantine periods prior to flights, such as those that astronauts have routinely done since the early days of human spaceflight, should continue to be used.[31] The journey from Earth to Mars could serve as another quarantine period. Anyone who shows signs of an infectious disease on the way should not be eligible to disembark upon arrival, or would need to spend time in another quarantine area before being released. In this way, quite a few of the infectious diseases we deal with on a regular basis would not be a problem for people on Mars.

Human diseases that are transmitted by insects such as mosquitoes could be kept out of Martian settlements by being careful about not allowing the insects as stowaways. That could cut out about 17 percent of all infectious diseases, including many of the most common and most deadly like malaria and yellow fever.

Another way that we could limit the diseases that make it to Mars would be to not bring their animal reservoirs. A large number of diseases, from rabies to influenza, are the result of an infection spilling over from an animal host to a human one.[32] For some diseases, like AIDS, there was a single spillover event or at least a small number of them. An evolutionary change in the virus that became HIV—human immunodeficiency virus—allowed it to bypass the human immune system. Prior to that, the virus was known as SIV—simian immunodeficiency virus—because it only infected simians, or primates. After the switch there was no looking back—HIV was now a human pathogen. In other cases, like influenza, ongoing spillovers occur between animal hosts—birds, in the case of influenza—and humans. The influenza virus is constantly evolving, allowing it to switch host species on a regular basis and causing us to have to get a new flu vaccine each year in order to keep up with the changes.

We could decide to not bring the animal hosts of these diseases with us to Mars. That would virtually eliminate any possibility of having these zoonotic diseases, as they are more formally called, in our Martian settlements. The host of every zoonotic disease is some type of bird or mammal.[33] Some are wild, like ducks, bats, or primates. Others are domestic, like chickens and dogs. If we choose to go this route, it will mean not having any animal livestock—yet another reason why all Martians will likely be vegan.[34] It would also mean that we would not have any pets. No dogs. No cats. This might Mars less attractive to many would-be settlers (myself included). But then again, it would save lives because there would be far fewer infectious diseases, including those we already know about and others that would undoubtedly emerge in the future if there were opportunities for new spillovers.

On the other hand, the ecosystems we create on Mars will contain far less biodiversity than any ecosystem on Earth. Not only will there be a smaller number of pathogens—that part is good—but there will be fewer plants and animals, and far fewer microbes in general. The microorganisms we bring with us, wittingly or unwittingly, will be a very small subset of the rich microbial diversity we are used to encountering. This could be problematic.

Since the 1950s there has been a widespread and very concerning rise in a variety of noncommunicable diseases, including asthma, Crohn's disease, type I diabetes, inflammatory bowel disorders, and many types of allergies. These conditions have existed for a long time, but the peculiar thing about their sudden and persistent rise was that the people most often suffering from them were among the most affluent. That's the opposite of how most disease outbreaks have happened in the past—it's usually the disadvantaged who are the hardest hit. In 1989, epidemiologist David Strachan noted the rising incidence of allergies among British children and proposed an explanation. He discovered that allergies were less common among kids who had older siblings. Perhaps being around an older sibling exposed children to common germs when they were young, and this somehow protected them from developing allergies, Strachan suggested.[35]

Strachan's idea became known as the hygiene hypothesis. It was soon expanded to explain many of the other conditions that were rising alongside allergies, all of which turned out to be connected in some way to chronic inflammation—the result of an overactive immune system, it was suggested. Advocates of the hygiene hypothesis suggested that the relative lack of infectious diseases among people living in developed nations was causing their immune systems to overreact when faced with even the most innocuous trigger, such as a peanut. This seemed to explain why affluent communities were more susceptible, because the poorer children were more likely to get exposed to an infectious disease that would help train their immune systems, teaching them to react with a measured response.

The latter half of the twentieth century also saw a trend toward people spending more time indoors, at least among those who could afford to. In addition to not being exposed to the pathogens that cause infectious diseases, affluent kids were becoming less likely to come into contact with a wide range of microbes that aren't harmful, or only mildly so. Throughout most of our evolutionary history, we have been intimately connected to the natural world. So intimately that our bodies were generally overflowing with microbes, from bacteria to worms. This is the context in which our immune systems evolved, and so when we find ourselves today in a world with far fewer of these "old friends," as our long-standing microbial partners have been called, our immune systems tend to overreact.

A team of Finnish researchers extended these ideas even further. Ilkka Hanski, Tari Haatela, and Leena von Hertzen noticed that the rise in autoimmune conditions also coincided with global declines in biodiversity of all types of organisms. They wondered whether people having less exposure to a rich array of living things—animals and plants as well as microbes—might be contributing to the autoimmune disorders. They dubbed this the biodiversity hypothesis.[36]

They found an interesting way to test their new idea. They knew that there were differences in the rates of allergies between people living on either side of the border between Finland and Russia in a region called Karelia. Prior to World War II, Karelia had been part of Finland, and the people in the area that now spanned two nations shared a common heritage. And yet the rates of allergies, asthma, and other autoimmune conditions were three to ten times lower for those on the Russian side than among the people just across the border in the Finland part of Karelia. For example, almost nobody on the Russian side was allergic to peanuts, whereas peanut allergies had become common on the Finnish side. But in other ways life was hard in Russian Karelia. People there lived in simple homes and raised most of their own food by keeping livestock and growing vegetables. The Finnish side had become much more Westernized, and, while rural, people there spent less time outdoors and were more likely to buy their food in stores than to grow it themselves.

Given the similarities between the people and the climate on either side of the border, Hanski, Haatela, and von Hertzen suspected that the dramatically different rates of autoimmune conditions were connected with differences in lifestyle. Specifically, they hypothesized that people in Russian Karelia were coming into contact with a greater number and diversity of microorganisms than people in Finnish Karelia.

They were right. As predicted, their data showed that the people on the Russian side had more microbes on their skin and in their noses than those on the Finnish side.[37] And bloodwork showed a more active immune response among those on the Finnish side, along with higher rates of allergies, eczema, and other autoimmune conditions associated with inflammation—consistent with the idea that less exposure to microbes leads to an overactive immune system.

The Karelia study supported the biodiversity hypothesis. It pointed toward a protective role of the diverse microbes that people encounter when they have more contact with the outdoors. But if going outdoors is good for priming our immune systems, then what would happen to people living in hermetically sealed habitats on Mars? They will rarely go outside, and even when they do, they will be wearing space suits, so they won't actually be exposed to anything living outside. But that doesn't even matter because, as far as we know, there is nothing living out there anyway. People on Mars will interact with only what lives inside the habitats with them, and that will be a small fraction of the microbial diversity that we are accustomed to here on Earth. This seems to be a recipe for developing a lot of inflammatory autoimmune conditions. So, allergies, asthma, type I diabetes, eczema, Crohn's disease, inflammatory bowel disorders, and other conditions could all become very common among Martians.

*

But there is more.

Here on Earth, when we encounter a microbe, whether it's a potentially dangerous pathogen or something harmless, our immune systems respond in several ways. One aspect of the response is to try to immediately neutralize and destroy the threat. This is called the innate immune system, and it works reasonably well most of the time. However, the innate immune system alone cannot always prevent an infection the first time we encounter a new type of microbe, which is why we are often susceptible to a new strain of flu, coronavirus, or other infectious disease. Another defense we have is called the adaptive immune system. It acts as a type of immunological memory in order to be able to respond more quickly if the same microbe ever shows up again. This involves the creation of antibodies, proteins that are made by our white blood cells and that remain there like a standing army, guarding and protecting us from the potential of any future invasion by the same microbe. Antibodies are what keep us from getting sick from the same bacteria or virus over and over again.

People on Mars will only develop antibodies to the small number of microbes they encounter in their sealed habitats. That means that any one of the vast number of microbes here on Earth that don't make it to Mars could potentially be dangerous for people born and raised on Mars. Traveling to Earth would be especially dangerous for Martians because they will have no immunity to many of our infectious diseases, much like the fictional aliens in H. G. Wells's *War of the Worlds*—except the aliens will be human. Even microbes that are harmless to people on Earth could be potentially dangerous to a Martian with no previous exposure to them.

The closest precedent we have for what this would be like is a genetic disease called severe combined immunodeficiency, or SCID. People with SCID cannot make antibodies, leaving them vulnerable to almost any microbe capable of causing a human infection. Consider the sad case of David Phillip Vetter, who was born at Texas Children's Hospital in Houston in 1971. His parents, Carol Ann Demaret and David J. Vetter, had already lost one child to SCID. When they learned that their son David would have the condition as well, they looked for a way to try to protect him for as long as possible. He was placed in a sterile enclosure immediately after birth. Inside, the baby could be kept safe from all microbes, but he could not leave the enclosure and could only interact with people, including his parents, through a clear plastic sheet.[38]

The enclosure kept him alive, but as David grew into a toddler and then a young boy, being confined inside it became increasingly challenging. His unusual case began to gain attention, and the media dubbed him the "bubble boy." David's doctors at Baylor College of Medicine reached out to their colleagues at Johnson Space Center, who designed a suit called the Mobile Biological Isolation System, modeled after a space suit.[39] They used their experience quarantining astronauts after their return from space, except that David's suit was even more challenging to build because the astronauts never had to worry about the inside of their suit being completely sterile. They came up with a design that required David to climb through a small tunnel when he got in and out of his suit to ensure that no microbes made it in.

It worked. For the first time in his life, at the age of six, David was finally able to go outside the hospital, even if it was only for about an hour at a time. It was also the first time his mother could hold David in her arms. For his eleventh birthday, David asked to go outside at night so he could see the stars. Tragically, David died at the age of twelve after a bone marrow transplant from his sister. Unbeknownst to the doctors, the bone marrow contained a latent type of herpes virus called Epstein-Barr virus—the same type that is commonly reactivated in astronauts during spaceflight. Without a functional immune system, David succumbed to a form of lymphoma caused by the virus. An article in the *Houston Chronicle* noted that "his mother was able to kiss him for the first time only when he came out of the bubble to die."[40]

As heart-wrenching as David's story is, a person who was born on Mars would experience similar challenges if they came to Earth. Vulnerability to microorganisms could be the most problematic aspect of traveling between planets. Martians will be most at risk when traveling to Earth, because Earth will have a great deal more microorganisms than Mars. But people from Earth who travel to Mars could also put Martians at risk because they are likely to carry at least some microbes on them, in their microbiome, that Martians have not yet encountered. Quarantines could also help limit the spread of infectious diseases to a certain extent by minimizing the transmission of microbes that are pathogens for Earth people, but they will not help to protect Martians from the great number of microbes that do not make Earth people sick. Likewise, immunizations against common Earth microbes could help to a certain extent, but would not be able to inoculate Martians against all of the vast diversity of Earth's microbial life.

I asked Sarah Wallace if this line of reasoning made sense to her. "Yeah, absolutely," she said. "You wouldn't be able to mount an immune response the way that someone on Earth could."

In some ways the situation is reminiscent of the contact between Europeans and Native Americans beginning in the fifteenth century. A great number of infectious diseases including smallpox, cholera, and measles were spread from Europeans to Native Americans. Because they had never been exposed to these diseases before, the Native Americans had no immunity to them. The result was devastating—by some estimates as many as 95 percent of the Native American population was killed in the following years. Similarly disastrous effects were seen among Pacific Islanders upon contact with the first Europeans.[41] However, although they were vulnerable to pathogens, the Native Americans and Pacific Islanders at least had the benefit of having been exposed to a diversity of common microbes prior to contact. Martians won't.

Making matters worse, founder effects will decrease the genetic diversity among Martian settlers. That will make them even more vulnerable to the spread of infectious disease, because a microbe that cracks the code for how to get past the immunological defenses of any one Martian would be well positioned to infect a great number of genetically similar Martians. Less diverse populations are always at greater risk from infectious diseases.

So, people born on Mars would get sick by coming to Earth, and contact with people from Earth could also make Martians sick. That means that, perhaps starting as soon as the first generation of people born on Mars, interactions between people from Earth and people from Mars would be risky. Physical contact between the two would be especially dangerous, particularly for the Martians. Intimate contact would be out of the question. The risk would likely increase with each new generation. In other words, don't expect any Martian-Earthling couples, and no mixed-planet babies.

If people from Earth and Mars never have children together, there will be no exchange of genes between their populations. Thinking back to studies of islands, it is gene flow—the exchange of genes between populations through reproduction—that prevents island populations from evolving enough differences to become new species. Without gene flow between them, people on Mars will accumulate genetic differences due to normal evolutionary processes. The effects of mutation, natural selection, and genetic drift will cause people on Mars to become increasingly different from people on Earth.

Perhaps more than any other factor, the risk of disease transmission may be the wedge that drives the separation between people on the two planets. It will, perhaps inevitably, cause the people on Mars to truly become Martians.

I asked Sarah Wallace for her personal opinion about whether the risk that future generations might not be able to return to Earth should stop us from building settlements on Mars. "Nothing should stop us from exploring," she replied. "If you take exploration out of the human imagination, what do we have? What good are we?"

Microbes and our bodies' responses to them will certainly pose a challenge to any attempts to keep Martians and Earthlings closely connected. But we are not entirely at the mercy of nature.

We still have a few more tools at our disposal.

8

THE NEXT HUMANS

Chris Mason is a man in a hurry.

"Sometimes walking from the subway to the lab takes too long, so I'll start running," he told me over breakfast at a bistro near his home in Brooklyn on a crisp autumn morning. "Just so I can get there faster. Not because I'm late for a meeting, just because it's taking too long to walk. . . . I'm the only one I know who runs to work to get there faster."

Mason is a professor of physiology and biophysics at Weill Cornell Medicine. At least that's his official title. He seems to be working on a hundred different projects all at once, ranging from tracking changes in the virus that causes COVID-19 to helping corals adapt to climate change.

"I love my job, and I feel like I need to get it done faster," he explained. Mason was in his mid-forties, with short, brown hair and a prominent jawline. If a movie were ever made about him, Matt Damon would be a good casting choice. He spoke in rapid fire as if his mouth were in a constant struggle to keep up with his brain. I had to remind him to take a few bites of his omelet.

The previous day I had visited his research group on the Upper East Side. The Mason Lab occupied four separate laboratories spanning three different buildings, and was still growing. Although they were pursuing a wide range of projects, a major focus of their work was on how the human genome and microbiome are affected by spaceflight.[1] Less than fifteen years after beginning his first faculty position, Mason's name seemed to show up on the list of authors of every other space genetics research paper I read.

In many ways, it sounded like he was doing exactly what he dreamed of as a kid. "I have the same goals I've had since I was thirteen," he told me. "Not much has fundamentally changed. I just finally have tools and resources to actually make progress on them."

As a boy growing up just south of Milwaukee, Wisconsin, in the 1980s, Mason was interested in many of the same things as other kids his age. "The two things I loved most were space and dinosaurs," he said. But unlike many other kids, Mason's interest in science only grew stronger as he got older, particularly his fascination with space. He got to go to Space Camp in third grade, and then again in fifth grade. That was exciting, but back home things were less interesting. School was especially boring. In middle school, he wasn't particularly engaged by most of his science classes—except for when he learned about genetics.

"The recipe for the book of life was starting to be revealed, but we could barely understand it," he recalled about the efforts to sequence the human genome. "But we knew the concept. There is a 3-billion-letter genetic code that defines creation of all the cell types and it's all there in the first cell. Once I learned that fact, with more detail, I was—and have been ever since—fascinated with genetics . . . just the fact that that code is there is the coolest fact in the universe—and it's different between species, and it adapts, and changes, it evolves . . . just learning that fact, that was like, that's it—I want to be a geneticist. I'm in eighth grade but I knew that's exactly what I wanted to do."

He went to college at the University of Wisconsin–Madison because it was one of only a handful of schools that offered an undergraduate degree in genetics, plus the in-state tuition rate made it a great deal. He got involved with some research projects, which he enjoyed, but they also left him wanting to pursue his own research questions. He went on to do a PhD at Yale, studying genetics in fruit flies, where he found himself on the cutting edge of some new ways of understanding how genes work. He was finally in charge of his own projects, and he found the process exhilarating.

"It was really fun to just see discovery—that moment where, when I analyzed the data, that I knew something that no one else in the universe knew at that moment," he recalled. "That joy of scientific discovery was my first kind of taste of it, which is also something I've loved ever since. The sense of discovering something completely novel that no one knows yet."

Figure 1.1 Ripples in the rocky bed of the Brazos River, in Central Texas, were identified as evidence for a 66-million-year-old tsunami. The timing and direction of the tsunami suggested an asteroid impact somewhere in the Gulf of Mexico. The outline of an impact crater created by a massive asteroid was later discovered in the Yucatán Peninsula near the town of Chicxulub. Source: Scott Solomon

Figure 1.2 SpaceX's Starship rocket sits on the launch pad at Starbase prior to its first test launch in April 2023. The upper stage and Super Heavy Booster were both designed to be reusable to reduce the cost of sending people and supplies to deep space. Stacked together, the booster and second stage were the largest rocket ever built. Source: Scott Solomon

Figure 1.3 The first detailed image of the surface of Mars came from NASA's Mariner 4 spacecraft during a flyby on July 14, 1965. The photos of the planet's surface were transmitted to Earth but it took time for the digital signal to be processed by a computer to form an image. While they waited, NASA engineers colored in strips of paper using different colors based on numbers from the camera that corresponded to the darkness recorded for each pixel. Source: NASA

Figure 1.4 NASA's Mars 2020 Perseverance Rover takes a selfie on July 23, 2024, using its robotic arm after exploring a region of Mars's Jezero Crater known as Bright Angel. At this site, the rover team found a rock with organic compounds and distinctive patterns that could have been formed by ancient microbial life. Earlier in its mission, the rover's MOXIE experiment demonstrated that the carbon dioxide in Mars's atmosphere can be split to make breathable oxygen, laying the groundwork for the future arrival of humans. Source: NASA

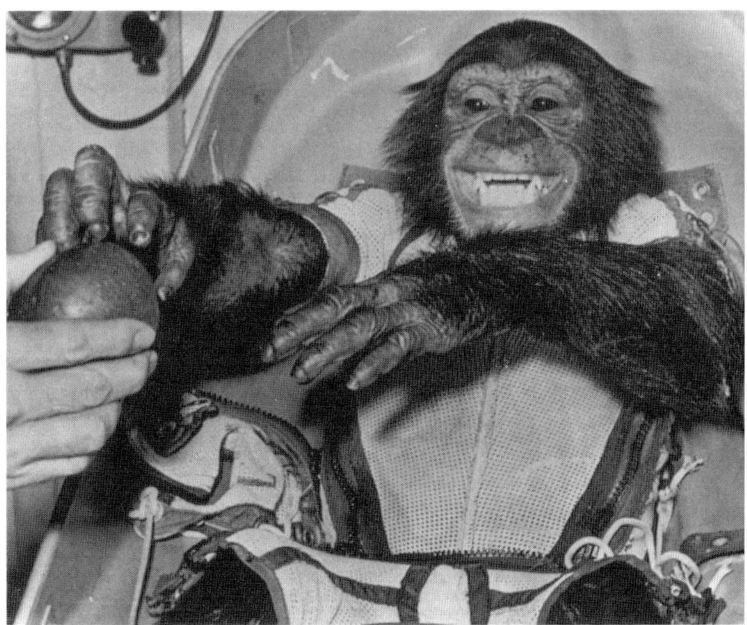

Figure 1.5 Ham, a three-year-old chimpanzee, flew into space on January 31, 1961, aboard a Mercury-Redstone rocket. One of forty chimpanzees trained at Holloman Air Force Base in New Mexico, Ham experienced up to seventeen g's followed by a brief period of weightlessness, but his performance in psychomotor tests during the flight suggested that being in space that would not incapacitate a human astronaut. Source: NASA

Figure 1.6 On April 12, Yuri Gagarin became the first human to travel into space. Gagarin's experience during his flight on a Vostok rocket launched from the Baikonur Cosmodrome in Kazakhstan laid to rest several concerns about how the conditions of space affect the human body, reporting that weightlessness was "no problem" and that it was possible to swallow. Source: NASA

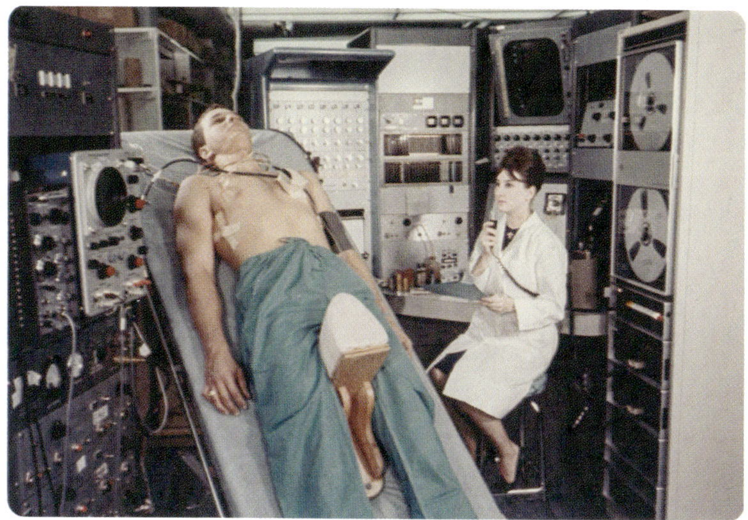

Figure 1.7 Margaret Ann Goldstein and an unidentified male volunteer during a NASA bedrest study in 1963. Bedrest was used to simulate the conditions of weightlessness, and early studies indicated that weakening of bones, muscle, and heart tissues could be dangerous for astronauts who were in space for more than two weeks. Source: Margaret Anne Goldstein

Figure 1.8 Astronaut James Irwin on the Moon in 1971. Irwin and David Scott spent nearly three days on the Moon, during which Irwin had a spacesuit malfunction and experienced a cardiac arrhythmia. Source: NASA

Figure 1.9 The crew of Soyuz 11, Georgi Dobrovolski, Viktor Patsayev, and Vladislav Volkov, became the first people to live in space when they spent twenty-four days on the Salyut space station in 1971. A sudden valve failure caused a leak in their capsule as it returned to Earth and all three men, who were not wearing pressure suits, died within minutes. Source: NASA

Figure 1.10 Astronauts Gerald Carr and William Pogue demonstrate weightlessness during Skylab 4. Skylab (1973–1974) led to major advances in the field of space medicine. Source: NASA

Figure 1.11 Astronauts John Glenn and Scott Parazynski perform biomedical research aboard space shuttle *Discovery* in 1998. At seventy-seven years old, Glenn became the oldest person to travel to space. The space shuttle era expanded the diversity of astronauts, increasing our understanding of how space affects men and women of different ages. Source: NASA

Figure 1.12 Cosmonaut Valeri Polyakov on the space station Mir. Polyakov set the record for the longest time in space, spending 437 consecutive days on Mir between January 8, 1994, and March 22, 1995. As a physician, a major motivation for Polyakov's long mission was to determine whether the human body could tolerate deep space missions. Upon his return, he announced, "We can fly to Mars!" Source: NASA

Figure 1.13 Astronaut Scott Kelly on the International Space Station during his year-long mission. Before, during, and after the mission, Kelly helped researchers collect extensive biomedical data to better understand how the human body responds to prolonged time in space. Similar data were collected from his twin brother, Mark Kelly, who remained on Earth. Source: NASA

Figure 1.14 A researcher prepares a sample in the target room at the NASA Space Radiation Laboratory. Located at Brookhaven National Lab, the lab includes a galactic cosmic ray simulator used to understand how radiation in space affects the body and mind. Source: NASA

Figure 1.15 Alexander Layendecker, Shawna Pandya, Egbert Edelbroek and Simon Dubé prepare for a presentation entitled "Sex in Space: Sex and Reproduction Beyond Earth," at SXSW 2023. Edelbroek announced that his company, SpaceBorn United, is aiming to develop technologies to enable in vitro fertilization (IVF) in space. Source: Scott Solomon

Figure 1.16 Japanese rice fish, or medaka, were the first animal to complete all stages of development from fertilization to birth in space. Source: Getty Images

Figure 1.17 Biosphere 2, located near Tucson, Arizona, was designed to simulate a self-contained space settlement. Beginning in 1991, a team of eight people was hermetically sealed inside. While they successfully completed the two-year mission, there were significant difficulties with the atmosphere, food supply, and crew dynamics. Source: Scott Solomon

Figure 1.18 Differences in shell shape between Galápagos tortoises from different islands were a clue that led Charles Darwin toward his theory of evolution by natural selection. Changes in body size among isolated populations of animals, such as the evolution of large body size in Galápagos tortoises, were first noted by biologist Bristol Foster in the 1960s, and became known as the Island Rule. Source: Scott Solomon

Figure 1.19 Evolutionary changes among future generations of Martians may include denser bones, shorter stature, and changes in skin pigmentation to protect against space radiation. Source: Joseph Ventura

Figure 1.20 Taputapuātea, on the island of Raiatea, in the Society Islands, is thought to be the launching point for ancient Polynesian voyages of discovery. Source: Scott Solomon

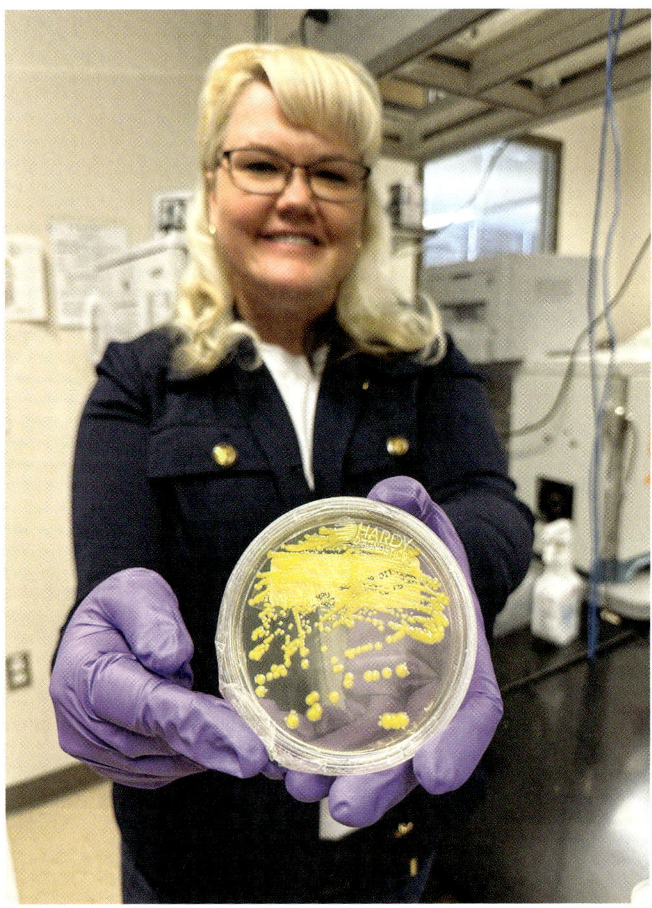

Figure 1.21 NASA Microbiologist Sarah Wallace shows a culture of bacteria used in studies monitoring microbial growth in the International Space Station. Studies suggest that some microbes rapidly evolve in space. Source: Scott Solomon

Figure 1.22 David Phillip Vetter was born with severe combined immunodeficiency (SCID), a genetic condition that compromised his immune system and required him to be isolated inside a sterile enclosure. NASA engineers built him a suit that allowed him to briefly go outside. Source: NASA

Figure 1.23 Geneticist Christopher Mason and his colleagues are using bioengineering techniques, including CRISPR gene editing, to help humans adapt to conditions beyond Earth. Source: Christopher Mason/Weill Cornell Medicine

Figure 1.24 Bina48, a humanoid robot designed by Hanson Robotics equipped with artificial intelligence, participated in a parabolic flight with Uplift Aerospace's commercial astronaut crew, Space+5. Crew member Mike Mongo later concluded that "AI is an evolved form of humanity." Source: Mike Mongo

Figure 1.25 The third integrated flight test of SpaceX's Starship and Super Heavy Booster (IFT-3) was the first time the upper stage successfully reached space. Source: Ryan Chylinski/Cosmic Perspective

In 2009 Mason got a faculty position at Weill Cornell Medicine and began building his own lab. He knew he needed to find funding sources, and it was an opportunity to take his research in a different direction. By then he had become broadly interested in how the body responds to extreme forms of stress, like cancer or other forms of disease. Going back to his earliest interests, he wondered whether the stress of being in space could be similar to other types of stress. He reached out to NASA to see if they would fund his work.

"The first grant I wrote was to NASA, as an unsolicited proposal," he said. Even more than money, what he needed was interesting samples to work on. Mason asked if there were biological samples from astronauts who had flown in space that his lab could analyze to look at the effects of spaceflight on the human genome and microbiome. He spoke with John Charles, who was still serving as Chief Scientist of NASA's Human Research Program, who told him that his proposal was interesting but that they did not yet have any samples to offer. His proposal was declined, but Charles hinted to Mason that they may have some opportunities coming up.

Indeed, just a few years later NASA announced that they were soliciting proposals for a study that would compare the health of an astronaut on a long duration space mission with a sibling on Earth—the Twins Study. Chris Mason sent in an updated version of his initial proposal, focused now on comparing samples collected from Scott Kelly during his year in space with those collected from his twin, Mark. This time, John Charles had good news for Mason—his lab was selected to lead the genetics, epigenetics, and microbiome aspects of the Twins Study.

Chris Mason was now at the forefront of research on space genetics. In the meantime, Mason had gotten involved in a student competition called iGEM, short for International Genetically Engineered Machines. The competition brought together teams of high school, undergraduate, and graduate students to work on ways to solve complex problems by modifying the genes of living organisms under the supervision of a faculty member. While the concept of genetic engineering had been around for decades, the tools for actually doing it were just coming into their own. When he was approached by a Cornell student about putting together an iGEM team, Mason saw an opportunity to apply genetic engineering to the challenges of human space exploration. He wrote a blog post in which he sketched out his ideas. They were ambitious and bold.

The basic premise was to take the process of biological adaptation to space into our own hands.

Settlement of space will lead to changes, one way or another. If we let it unfold naturally, evolution will take its course and people on Mars will gradually become better adapted to the conditions there through mutation and natural selection. Founder effects and genetic drift will cause random changes in Martians and reduce their genetic diversity. Enforced quarantines due to the risk of spreading infectious diseases could accelerate the speciation process. But these natural evolutionary processes will be relatively slow and, to put it mildly, quite unpleasant. What if we could accelerate the process of adaptation and minimize the human suffering that it would otherwise entail?

Mason thought that we could—and his blog post gave an overview of how it could be done. Robert Prior, an editor at the MIT Press, read Mason's blog post and encouraged him to expand it into a book, which was published in 2021 with the title *The Next 500 Years: Engineering Life to Reach New Worlds.*[2]

In the book, Mason laid out his argument in detail. "One possibility is we simply allow evolution to gradually select for characteristics required to survive on these new planets," he wrote. "This is basically the 'sink or swim' approach to life's survival, except with no lifeguards and bricks tied to your feet."

There is an alternative, he argued. "Our second option to enable Earth's life to live on other planets is to preemptively direct this genetic process, so that the life we send is already capable of surviving in its new home. More complex, yes—but also more humane."

The rest of the book describes Mason's plan for how exactly to do that. A key aspect of Mason's vision for genetically engineering people for life beyond Earth is to draw on the vast library of genetic adaptations that exist in the genomes of other living things. "Integrating one species' cells with those from another species' evolutionary history and learning these 'genetic lessons' can act as our guide to protect human cells and the cells of other species which will inevitably accompany us on our journey away from Earth," he wrote.

*

The basic idea of acquiring new abilities by taking DNA from one organism and putting it into another has existed since the 1970s. In 1972, biochemist Paul Berg became the first person to do this when he copied a short piece of DNA from a virus that attacks bacteria into another kind of virus that attacks monkeys.[3] The following year, Herbert Boyer and Stanley Cohen applied this "gene splicing" technique to insert genes from one species of bacteria into another. They found that the inserted gene was still there in subsequent generations as the bacteria divided. Going one step further, they then spliced genes from a frog into a bacteria, and found that the frog genes became a permanent addition to the bacteria's genome.[4]

This was the dawn of a revolution in biotechnology. Recombinant DNA—meaning DNA copied from one organism and pasted into another—could be used for an incredible number of things, from producing life-saving medications like insulin to more whimsical applications like making glow-in-the-dark cats and goldfish.[5] But it was also the beginning of an era in which humans could directly control the evolution of any species by manipulating their DNA. Mason saw the potential to take genes from organisms that are naturally well adapted to harsh conditions and insert them into human cells to help prepare people for the hazards beyond Earth.

One candidate for such a hardy creature is the water bear, or tardigrade. Tardigrades are distant relatives of insects but have a unique appearance. They are barely visible to the naked eye, but under a microscope they look something like a tiny gummy bear with eight chubby little legs and a mouth shaped like a nozzle. They thrive in moisture, but their adaptability allows them to live almost anywhere, from the sea to the soil in your backyard. One of the ways they are able to live in such a wide range of habitats is by being able to tolerate long periods of bad conditions—say, a drought—by essentially shriveling up. In their dehydrated state they are almost invincible, which is what has drawn the attention of biologists interested in life in outer space.

In 2007 an experiment by the European Space Agency launched about 3,000 tardigrades aboard a Soyuz rocket into low Earth orbit for twelve days.[6] Once in orbit, a compartment on the outside of the capsule where the tardigrades were housed was opened, which, in addition to microgravity, exposed them to the vacuum of space with unfiltered levels of solar and galactic cosmic radiation. Remarkably, upon their return to Earth, many of the tardigrades had survived by being in their dehydrated state—the same strategy they use to wait out droughts and other temporary stresses. In fact, there was no difference in survival between the tardigrades exposed to the vacuum of space and the tardigrades in the experimental control group, which were kept under normal conditions in the laboratory. After returning to Earth and being rehydrated the tardigrades that had flown in space were even able to lay eggs.

In 2016, a team of Japanese researchers led by Takekazu Kunieda and Atsushi Toyoda sequenced the genome of one particularly hardy species of tardigrade.[7] In the process, they discovered that the tardigrades make a protein that helps them to survive while in their dehydrated state. They named the protein "damage suppressor," abbreviated Dsup. The researchers then took a major leap: they took the gene for Dsup from the tardigrade genome and temporarily spliced it into human cells. To be clear, the human cells were growing in a laboratory, not a human body. Nevertheless, they found that with the tardigrade gene inserted, the human cells could produce Dsup. And—most significantly—when they exposed the human cells making Dsup to radiation in the form of X-rays, the cells were less damaged and better able to grow than normal human cells.

Chris Mason's lab began working to further improve the abilities of human cells to withstand the harsh conditions of space by splicing in genes from tardigrades and other organisms able to survive in extreme environments. He sees this as the beginning of an era in which human cells can be imbued with a great variety of abilities. He predicts that by the year 2040, "genes from all organisms will become a playground for creating and making new functions in human cells."

*

The notion that the diversity of life on Earth represents a genetic "playground" for us to draw from seems exciting, but is also fraught with risks, ranging from the biological to the ethical. Indeed, after the first demonstrations of gene splicing by Berg, Boyer, and Cohen, researchers quickly recognized the double-edged-sword nature of the new technology. A voluntary moratorium was called for by Berg, Boyer, and others on the use of recombinant DNA. A conference was held in 1975 at Asilomar Beach in California, in which many of the leading researchers came together in an attempt to develop a set of guidelines for its use. The meeting, which has variously been called "Woodstock for molecular biology" and "the Pandora's box conference" was in part an attempt by researchers to come up with their own limits on the use of genetic technology in the hopes that doing so would prevent government regulations.[8] Indeed, after four days of meetings, the researchers agreed to lift their own self-imposed moratorium on the use of recombinant DNA, albeit with some guardrails intended to prevent what they saw as its most potentially dangerous uses.

It seemed plausible that some genetic diseases could be cured with recombinant DNA by swapping out the section of DNA responsible for the condition with DNA from a healthy donor. The first attempt at this so-called gene therapy was performed in 1980 by Martin Cline, a geneticist at UCLA. Two patients with a blood disease called beta-thalassemia were treated using an approach that Cline had tested in mice, with only modest results. He went ahead and attempted it on humans anyway.[9] However, Cline had jumped the gun. He had not followed all of the necessary procedures to get permission to perform the experimental treatment in human subjects, and was reprimanded. Anyway, the gene therapy did not have any effect on the patients.

The first success came ten years later with a child named Ashanthi DeSilva. At the age of two she had been diagnosed with SCID—the same condition as the "bubble boy," David Vetter, that results in a nonfunctional adaptive immune system. In 1990, when DeSilva was four years old, she was given a fully approved experimental gene therapy to replace the cells in her bone marrow that cause the condition. It worked. With the modified genes, DeSilva's immune system began functioning well enough for her to go outside, attend school with other kids, and lead a normal life.[10] As an adult, she became a genetic counselor and an advocate for people with immune deficiencies.

More gene therapies soon followed, but not all were as successful. In 1999, an eighteen-year-old named Jesse Gelsinger died just four days after being given an experimental gene therapy meant to treat his liver disease. He had a relatively mild condition that caused his blood to accumulate ammonia, but was able to manage it with medications and a careful diet. Gelsinger was part of a clinical trial at the University of Pennsylvania that was intended to help treat young children with severe forms of his condition. Tragically, Gelsinger's immune system had a major reaction to the virus that was delivering the modified genes, causing a high fever followed by progressive organ failure. His death put a pause on many gene therapy trials until safer methods could be developed.[11]

The problem was not really with the modifications to the genes themselves but with the way in which the modified genes were being delivered. Ever since Berg, Boyer, and Cohen had developed gene splicing in the early 1970s, the approach had been to replace one piece of DNA with another that had the desired sequence. But this "cut-and-paste" technique meant the replacement DNA had to be delivered to the chromosomes inside a person's cells. This could be done very effectively with a virus acting as the delivery vehicle, since viruses are adept at penetrating cells and manipulating their DNA. But on occasion, the body can mount an immune response to the virus, which is precisely what killed Jesse Gelsinger.

*

A major breakthrough came with the discovery of a way to edit DNA directly. It happened, like many discoveries in science, in a roundabout way. In 1990 Francisco Mojica was a graduate student at the University of Alicante in Spain, studying a type of single-celled microbe called archaea. Many archaea live in environments that would be extreme for other organisms. Indeed, Mojica's archaea were found in extremely salty ponds. After sequencing some of their DNA in the hopes of learning how they tolerate so much salt, he found something unexpected and odd. In between sections that looked to him like normal DNA, with the usual combination of all four DNA bases A, T, C, and G, were sections that kept repeating the same bases. Even more strange, these repetitive sections were also palindromes, meaning that they could be read the same forward and backward. He found fourteen of these sequences clustered together at regular intervals around otherwise normal sequences of DNA.

Puzzled, Mojica searched the scientific literature for anything similar in other organisms. He only found one—from a sequence of *Escherichia coli* bacteria published by a Japanese researcher. Archaea and *E. coli* are completely different types of microorganisms, belonging to two distinct domains of life. Nevertheless, they both seemed to share the same peculiar cluster of repetitive DNA sequences. He published his results, unsure of what the function of the sequences might be. He would later give them a cumbersome name with a catchy acronym—clustered regularly interspaced short palindromic repeats, or CRISPR.

Soon, CRISPR sequences were found in a wide range of other microorganisms. Researchers in the dairy industry found them in the bacteria that ferment milk into cheese and yogurt. Intriguingly, they noticed that new CRISPR sequences appeared in the dairy bacteria after an attack by viruses—and that the CRISPR sequences matched sequences from the viruses' genomes. What's more, the bacteria with the new CRISPR sequences were no longer vulnerable to attack from the same virus. The CRISPR sequences were acting as a type of immune response by the bacteria: the bacteria were learning to recognize the virus so that they could defend against it in the future.

The mechanism for how CRISPR works was figured out by a team of researchers led by biochemists Jennifer Doudna and Emmanuel Charpentier at UC Berkeley. They discovered that CRISPR works with the help of proteins, called Cas—short for CRISPR-associated proteins. Cas proteins, like one they called Cas9, cut DNA like a molecular scalpel. Bacteria use CRISPR-Cas9 to recognize the unique DNA of a particular virus and then chop it up to destroy it. But what Doudna and Charpentier also found is that they could control what sequence of DNA is being targeted. It didn't have to be DNA from a virus. It could be DNA from any living thing. If the DNA is inside a living cell, the cell's machinery will naturally repair the damage. But the most exciting part of all was that Doudna and Charpentier found that they could manipulate the repair process so that a stretch of DNA could be cut out and replaced with any sequence they wanted. In other words, it was programmable.

"In the history of science, there are few real eureka moments, but this came pretty close," wrote Doudna biographer Walter Isaacson about the breakthrough.[12] Unlike the copy-and-paste gene-splicing approach, CRISPR would be able to alter DNA sequences by making precise, deliberate edits to an organism's genes. "In short, they realized that they had developed a means to rewrite the code of life," wrote Isaacson.

Doudna and Charpentier published their landmark findings in 2012.[13] Over the next few years, a race was on to figure out how to use CRISPR to edit human genes. It got spicy. George Church, a geneticist at Harvard with a reputation for being something of a maverick, was collaborating with Doudna but also working on his own approach to apply CRISPR to human genes. Meanwhile, Church's former student Feng Zhang had started his own lab at the Broad Institute, an elite biomedical and genomics research organization jointly run by Harvard and MIT. Zhang had recruited one of Church's current students to join him to work on developing CRISPR for human gene editing.

All of these groups knew they needed to work quickly if they wanted to be the first to claim the prize of editing human genes with CRISPR. A lot was at stake. In academic research, being first on a major discovery can make a person's career. But there was more than academic fame and glory on the line. There was also the possibility of major prizes, as well as patents and the potential to make some serious cash. In the end, Church and Zheng both submitted papers to the journal *Science* close enough in time that they were published simultaneously on January 3, 2013. Doudna's group published theirs later that month in another journal.[14] All of them had developed tools that could be used to edit human genes using CRISPR, although the Nobel Prize for its discovery, awarded in 2020, was given to Doudna and Charpentier.

Clinical trials were underway just a few years later to test whether CRISPR could be used to treat conditions ranging from diabetes and blood disorders to certain forms of cardiovascular disease and cancer. In 2023, the first two CRISPR-based treatments were approved in the United States. One was for sickle-cell disease, a condition that causes red blood cells to change their shape, leading to potentially deadly blood clots. It affects more than seven million people worldwide and kills more than 300,000 people every year. The other approved treatment was for another blood disorder, beta-thalassemia, the same condition that was the target of the very first attempt at gene therapy. Thalassemias reduce the ability of red blood cells to transport oxygen and can cause anemia as well as cardiovascular, endocrine, and respiratory disorders, among other complications. More than 11,000 deaths occur each year from thalassemias, and they affect more than a million people worldwide.[15] Because they can repair the genetic defect that causes both sickle-cell disease and beta-thalassemia, the CRISPR-based treatments have the potential to completely cure people who would otherwise suffer from these diseases.

There is a catch, however. While the hope is that patients receiving the CRISPR treatments will be fully rid of their diseases, the gene-editing approaches that have been approved so far would not prevent any of the patients' children from inheriting their parents' diseases. The genetic changes are only made to DNA in somatic cells—the cells of the body that are not involved in making sperm or eggs. For their children to be cured, they would need to undergo the same treatment as their parents. The same would be true of every subsequent generation.

The alternative would be to make edits to cells in such a way that they affect not only the somatic cells, but also the cells in the germline—those that become eggs, sperm, and eventually embryos and then babies. Germline gene editing is possible, although it crosses a line that some believe should not be crossed. The reason is that any edits made to germline cells will affect all the descendants of the individual receiving the treatment, for countless generations. This raises new types of ethical questions. It is one thing to perform a procedure on a living person, who can be educated about the potential risks and benefits and who can give their informed consent. Is it ethical to make decisions that will directly affect future generations who will not have any choice in the matter?

*

Where some saw risk, others saw opportunity. Even before CRISPR, when gene editing was mostly a hypothetical idea, there were some who thought germline gene editing would be an overwhelmingly positive development. At a conference on the topic held at UCLA in 1998, biologist Lee Silver argued that the ability of a person to edit their DNA—including that of their germline—is a matter of individual freedom and should therefore not be restricted. James Watson, the co-discoverer of the DNA double helix structure, agreed. "It seems obvious that germline therapy will be much more successful than somatic-cell edits," he said. Watson's own experience with heritable genetic disorders (his son suffers from schizophrenia) made this a personal as well as a professional matter. Watson didn't see the germline to be a red line. "The biggest ethical problem we have is not using our knowledge and not having the guts to go ahead and try to help someone," he said.[16]

Watson was asked whether making such heritable changes would amount to playing God. He replied, "if scientists don't play God, who will?"

As the CRISPR technology she helped to develop quickly moved from idea to reality, Jennifer Doudna had concerns about the potential for its misuse, particularly regarding changes to the human germline. She organized a gathering of leading scientists in 2015 to develop some recommendations on the use of germline gene editing. The meeting included several people, including Paul Berg, who had participated in the Asilomar conference, which was precisely what she hoped to emulate. Much like Asilomar, the gathering took place at an idyllic California retreat, in this case in Napa Valley. Doudna, Berg, and the other participants concluded that germline gene editing may prove to be an important tool, but they felt that it should not be used yet. They published a letter in *Science* explaining their reasons.[17]

One of the risks, they explained, was the potential for off-target effects—changes to DNA sequences other than those being targeted. Because of their concerns about these "unintended consequences" and the potential for a "slippery slope," they called for a temporary global moratorium on germline gene editing in humans. They were not asking for a permanent ban, but what they called a "prudent path forward."

In December 2015 an International Summit on Human Gene Editing was held in Washington, DC. Many of the leading CRISPR researchers were there—Jennifer Doudna, Emmanuel Charpentier, George Church, Feng Zheng—as well as representatives from universities and research organizations around the world.

"We could be on the cusp of a new era in human history," said David Baltimore, a geneticist from CalTech who was chair of the organizing committee, in his opening remarks. Baltimore had attended the Napa conference and had been one of the organizers of the Asilomar meeting. "Today, we sense that we are close to being able to alter human heredity. Now we must face the questions that arise. How, if at all, do we as a society want to use this capability? This is the question that has motivated this meeting," he said.[18]

The summit concluded with recommendations similar to those that came out of the Napa meeting. There needs to be more discussion of what uses of germline gene editing are appropriate, their report suggested. They listed a series of steps that should be followed for any gene editing of human germline cells. But they stopped short of calling for a total ban.[19]

Nevertheless, it came as a shock when, in 2018, the birth of the first human babies with germline gene edits using CRISPR was announced by Chinese biologist He Jiankui.[20] In his rush to be the first to perform the procedure, He had not followed the internationally recommended guidelines and had not applied for permission to conduct the clinical trial in China where the procedure was done. He also neglected to inform his university of what he was doing.

The twin baby girls, named Lulu and Nana, were apparently healthy. He had used CRISPR and its companion protein, Cas9, to edit a gene called CCR5. The specific edits he made changed the sequence of DNA in the CCR5 gene by removing a stretch of thirty-two bases, resulting in a sequence for the gene identical to what some people naturally have. Those with the deletion are less vulnerable to HIV, the virus that causes AIDS. The babies' father was HIV-positive but their mother was not, so He reasoned that the gene edits would help to protect the children from becoming infected by HIV. If the girls ever grow up and have their own children, they would inherit the edited genes.

News headlines announced the arrival of the world's first "CRISPR babies." But rather than being hailed as a hero, as he had hoped, He Jiankui was widely criticized. Geneticists pointed out that the change He had made in the CCR5 gene has a downside—while it should decrease the chances of the twin girls contracting HIV, it could make them more susceptible to other diseases, like West Nile virus. Was the trade-off worthwhile? There are other ways to protect against HIV, so the gene editing technique did not seem to meet the established criteria of being medically necessary. A Chinese court found He guilty of conducting an illegal medical practice. He was sent to prison and banned from performing any future work on human reproduction.[21]

Yet, in the next few years, global events would lend a new perspective on He Jiankui's actions. Walter Isaacson noted that "in the wake of the 2020 coronavirus pandemic, the idea of editing our genes to make us immune to virus attacks began to seem a bit less appalling and a bit more appealing."[22] In the aftermath of the CRISPR babies scandal, governments around the world scrambled to clarify the legality of germline gene editing in their nations. Some, like the United States, adopted more restrictions, while others, like Japan, softened theirs. While it seemed clear to most that He had acted prematurely and irresponsibly, his work did show that germline gene editing was possible in humans. The question was no longer about if it could be done; it was more about how and when.

*

Changing the germline means we are, whether we realize it or not, controlling the future of evolution. Yet while the techniques of gene editing are new, the idea that we humans have the ability to guide evolution is not. As Chris Mason pointed out, we have been doing so for millennia through the practice of selective breeding in agriculture and the domestication of animals.

"While controlling the evolution of the past, present, and future seems scary and wrought with incredible hubris, the reality is that we already have been engineering and modifying species and the environment around us, except previously we were doing so by accident with no foresight," Mason wrote.[23] "Now, finally it can be done with a sense of responsibility and purpose."

Yet the idea of purposefully controlling the evolution of our own species has a dark history.

In 1883, Francis Galton proposed improving our species through selective breeding in much the same way we do for animals, which he described as "the science of improving stock."[24] Applying this to humans, he argued, would "give to the more suitable races or strains of blood a better chance of prevailing speedily over the less suitable." Galton was Charles Darwin's cousin, and he was attempting to apply Darwin's ideas on evolution to what he saw as the degradation of humanity through his overtly racist lens. Galton had attempted to figure out how heredity works by studying genealogies of both people and animals, such as dogs. Among his investigations were the first studies of twins and an attempt to determine what physical characteristics criminals had in common so that they could be recognized prior to committing any crimes. Based on his observations, Galton thought it would be possible to make the characteristics he considered positive—like good health, intelligence, and responsibility—more common in society by encouraging marriages between people from families with a history of these characteristics. He called this idea "eugenics."

As Galton's ideas spread, they also evolved. In addition to encouraging the breeding of people with supposedly good characteristics, some sought to achieve similar results by preventing the reproduction of people with traits they considered undesirable. The first government to enact laws based on eugenics was the state of Indiana in 1907, followed soon after by thirty-one other US states. The laws included forced sterilization for people labeled "criminals, idiots, imbeciles, and rapists." The issue was brought before the Supreme Court in 1927. The question was whether a twenty-one-year-old woman named Carrie Buck could be surgically sterilized because she had been labeled an "imbecile." A central argument in the case was the determination that Carrie Buck's mother and baby were also "imbeciles," which the prosecution argued made it clear that the condition was hereditary. In an 8–1 ruling, the Court determined that forced sterilization was indeed legal.[25]

"It is better for all the world," they wrote, "if instead of waiting to execute degenerate offspring for crime, or to let them starve for their imbecility, society can prevent those who are manifestly unfit from continuing their kind . . . three generations of imbeciles is enough."

In Germany, the Nazi party modeled their policies on the American eugenics laws.[26] They passed a law in 1933 that mandated surgical sterilization for anyone they determined to be carrying a "hereditary disease."

But sterilization was just the first step. Soon, the Nazi efforts of "racial hygiene" would include murder and genocide.

The atrocities committed in the name of eugenics in the first half of the twentieth century were based not only on prejudiced and racist views, but also on flawed science. We now know that there is little if any genetic basis for the traits that proponents of eugenics sought to control. Still, any discussion of manipulating the future of human evolution has to consider the flaws inherent in previous attempts to do so, as well as the ways in which those efforts were perverted and abused.

In her book about the discovery and potential uses of CRISPR gene editing, *A Crack in Creation*, Jennifer Doudna addressed the question of eugenics directly.[27] "This fear—that gene editing will exacerbate existing prejudices against people who fall outside a narrow range of genetic norms—underlies the association that numerous writers have made between germline editing and eugenics," she wrote. "Given the deplorable history humans have when it comes to programs aimed at improving our species' gene pool, perhaps it's not surprising that CRISPR's potential to endow individuals with healthier genes earns it comparisons to these sad chapters in our past."

But Doudna argued that germline gene editing is not fundamentally different from other practices commonly used in modern medicine, including the standard practice of IVF, in which embryos are screened for genetic diseases before being implanted in the uterus.

"Eugenics, as most people remember it today, was certainly reprehensible, but the odds are miniscule that we'll see anything similar happen with gene editing," Doudna wrote. The reason, she explained, is that governments are unlikely to mandate gene editing. The way she sees it, "germline editing would remain a private decision for individual parents to make for their own children, not a decision for bureaucrats to make for the population at large."

She went on to argue that germline gene editing should be allowed under the right circumstances, which include making it both safe and equitable. "If we can do these things," she wrote, "if we can walk the narrow line between prohibiting CRISPR to the detriment of certain individuals' health and overusing it and subverting our society's values—we will be able to use this technology in a way that is unequivocally good."

*

Questions about the ethical use of gene editing become more complex when considering humans on other planets. Would it be ethical to change a gene to make a person traveling to Mars better able to tolerate lower gravity or higher radiation? What about for a child born on Mars? Could genome editing make it easier to allow people to move safely between planets, for example, by altering their immune systems?

Chris Mason sees gene editing in the context of space settlement as a moral imperative. "Sending any Earth-evolved organism to another planet would result in almost certain death, which represents the sad, evolutionary 'good luck' plan," he wrote.[28] "To save life, we will need to engineer it." To Mason, using genetic engineering is not only about improving life; it's about preserving it. "When given the choice between engineering life or facing inevitable death, there is clearly only one path. The right thing to do, in order to survive extinction, is to engineer at a genetic, cellular, planetary, and interstellar scale," he wrote.

Mason's reasoning is based on an ethical philosophy he calls deontogenics. According to this way of thinking, as a species that is aware of the possibility of our own extinction and that of other species, we have an ethical obligation to try to prevent that from happening. He refers to the collection of humans and other species from Earth that we can bring with us as a single entity he calls the "metaspecies."

"Any act that consciously preserves the existence of life's molecules . . . across time is ethical. Anything that does not is unethical," Mason wrote.

He went on to elaborate what this philosophy entails, arguing that our survival depends on our ability to make it to other solar systems and that "the needs of the metaspecies and conservation of their responsibilities may supercede individuals' wants or needs." Failure to act—in other words, not taking action to preserve life by figuring out how to make it capable of thriving on other planets—would be unethical under Mason's deontogenic philosophy. "It is equivalent to sitting on train tracks, knowing that a large train is coming down the tracks to destroy you, and just waiting for it," he wrote.

As we finished our breakfast, Mason told me about how he developed his deontogenics philosophy. Basically, he takes the long view.

"All ethical questions become crystal clear through the lens of a billion years," he explained. "I always try to think, will this question matter in a billion years?"

With this framework in mind, Mason and his research team are pressing forward on genetically engineering human cells to make them better adapted for conditions beyond Earth. They have had some success with getting human cells to produce the Dsup protein that helps tardigrades survive in space. So far, their work involves only human cells being grown in a lab, but he hopes that will soon change. "I'd say human trials are ten years away," he told me.

A list of other genes that could be modified to help people to deal with life on Mars and elsewhere has been identified by George Church, Chris Mason, and colleagues at Harvard's Consortium for Space Genetics.[29] They include genes that influence bone density, muscle tone, radiation resistance, and even pain tolerance. In part, the list comes from studies of existing genetic variation within people alive today. It also comes from organisms capable of living in extreme environments, like tardigrades and others.

One particularly hardy species of bacteria was first discovered in the 1950s in a can of meat that had been exposed to a whopping dose of 5 million millisieverts of radiation. The goal was to see if radiation could be used as a way to sterilize canned foods to make them safe to eat. Yet the bacteria were still alive. The researchers identified them as belonging to the genus *Deinococcus* and named the species *radiodurans* in reference to their remarkable ability to endure such high radiation exposure. Even tougher bacteria, like the appropriately named *Thermococcus gammatolerans*, have been found in the water used to cool nuclear power plants.[30] The genetic basis of these species' abilities to withstand radiation is being investigated by Mason and his colleagues for their potential use in engineering life beyond Earth.

Another approach Mason is researching is to genetically engineer genes in bacteria and other microbes in our microbiome to produce useful products, including Dsup. This way, no changes to human cells would be required, but people might still be able to reap the benefits if the substances made by microbes are active within the human body. They already have some microbes in the lab that seem capable, he told me, but so far they have not tested whether the microbes would work the same way when living in humans.

"It's still a few years before we do a trial like that," Mason said.

*

So, we can edit our genes or those of our microbial partners. But there is yet another way that genetic technology may facilitate the human migration into space: by creating genes that do not yet exist.

In the first decades of the twenty-first century the field of synthetic biology emerged, with the goal of creating new functions for living things using genetic tools. Scientists at the J. Craig Venter Institute created the first complete synthetic genome of a simple organism, a type of bacteria, in 2010. Work has since been underway to create synthetic genomes for more complex organisms. Eventually, synthetic human genomes might be possible.[31]

One idea, suggested to me by biologist Tiffany Vora, is to create synthetic portions of a human genome. For example, while humans normally have twenty-three pairs of chromosomes, one or more new chromosomes could be made that could augment our existing genome.

"The idea that we're going to find all the mutations we need in Earth's situations—I don't believe it, because we're fundamentally looking for a non-Earth context," Vora told me. The advantage of this approach is that the existing genome could be left untouched. "If you can make really long artificial chromosomes, then you don't have to change the person—you just give them a patch, essentially."

Changes in the number of chromosomes happen routinely in evolution. But when they do, it can trigger speciation because individuals with unequal numbers of chromosomes are often unable to mate and produce healthy offspring. This is what keeps horses and donkeys distinct, according to Ernst Mayr's definition of species.[32]

In fact, a change in the number of chromosomes may have been an important part of our own evolutionary history. All living great apes—including chimpanzees, gorillas, and orangutans—have twenty-four pairs of chromosomes. Our species has twenty-three pairs, and so did our extinct relatives the Neanderthals and Denisovans. What seems to have happened is that, about a million years ago, parts of two different chromosomes fused together to create one large chromosome. That event may have played a role in separating our ancestral lineage from the others alive at the time while still allowing reproduction between *Homo sapiens*, Neanderthals, and Denisovans.[33]

This raises the possibility that future humans with additional synthetic chromosomes may be genetically incompatible with people without them. If used for space settlement, this could be yet another force driving a wedge between humans from Earth and humans living elsewhere. Adapting to life in space may require genetic engineering, but engineering people for space might also contribute to a split in humanity. At some point, people may have to choose whether to prioritize adaptation for life on other planets or the ability to maintain human beings as one single species. It might not be possible to achieve both.

*

Other ideas for how to use technology to help people adapt to life beyond Earth include enhancing our bodies with mechanical, electronic, or robotic components. We are already accustomed to wearing glasses, using hearing aids, prosthetic limbs, artificial hearts, and many other devices to improve human health and well-being. Brain–computer interfaces can be added to the list. In 2004, Matthew Nagle became the first person to have a device implanted in his head that connected his brain to a computer. Nagle became paralyzed below the neck after being stabbed in the back with a knife. But the brain implant allowed him to control a computer mouse with his mind. By 2012, a new type of brain implant allowed fifty-five-year-old Cathy Hutchinson, who had been paralyzed for fourteen years, to control a robotic arm well enough that she could drink on her own. The 2014 World Cup was literally kicked off by a twenty-nine-year-old paraplegic Brazilian man named Juliano Pinto, who used a brain–computer interface to control a mechanical exoskeleton, allowing him to kick a soccer ball.[34]

Numerous private companies working on brain–computer interfaces have recently emerged, suggesting the technology is maturing. In 2024, Neuralink—another company owned by Elon Musk—implanted its first experimental device in a human patient.[35] As the technology improves, brain–computer interfaces will allow better control of artificial limbs and exoskeletons as well as other devices such as vehicles, robots, and more.

These technologies could certainly be helpful for life on other planets. Connecting the brain to devices that enhance the senses could give people the ability to see or hear in ways that our eyes and ears cannot do on their own. In conjunction with advances in robotics and artificial intelligence, this technology could lead to abilities reminiscent of the Marvel comic hero Iron Man. Imagine a Mars rover, with all of its sophisticated tools and machinery, controlled entirely by the human mind. Now imagine that you *are* the rover. Humans with these enhanced abilities could become the most capable and best adapted Martians.

In fact, human–machine hybrids—commonly known as cyborgs—may be our best option for deep space exploration. Indeed, the term "cyborg" was first coined in a 1960 article about the future of human spaceflight by psychiatrist Nathan S. Kline and inventor Manfred Clynes.[36] "If man in space . . . must continuously be checking on things and making adjustments merely in order to keep himself alive, he becomes a slave to the machine," they wrote. "The purpose of the Cyborg . . . is to provide an organizational system in which such robot-like problems are taken care of automatically and unconsciously, leaving man free to explore, to create, to think, and to feel."

A photo in the article labeled "one of the first cyborgs" showed a rat with a syringe attached to it that periodically injects medications into the rodent's bloodstream. Kline and Clynes suggested that a similar device could be adapted for use by astronauts, which could automatically inject medicines to help with changes in radiation, gravity, and temperature, or other space-related challenges. They imagined that other types of implantable devices could be developed, including machines to help with breathing by converting carbon dioxide into breathable oxygen. "Such a system, operating either on solar or nuclear energy, would replace the lung, making breathing, as we know it, unnecessary," they wrote.

*

Perhaps the future of humans in space will indeed involve merging with machines. If it does, however, I would not expect it to cause a fundamental change in our evolution the way that germline genetic engineering will. The reason is that cybernetic technologies, whether artificial lungs or brain–computer interfaces, are not heritable. They will need to be installed or implanted in each generation. They might help us to survive, or even reproduce—no word yet on what that type of cyborg would entail—but making a change to one generation will not automatically change the characteristics of future generations. They might just make future generations more likely to exist.

At least, that was how I thought about it until I met astronaut teacher Mike Mongo at the New Worlds space conference in Austin in 2023.

Mike is one of those larger-than-life personalities. If he is in a room, you'll know it. He is tall, boisterous, energetic, and outgoing. But his trademark is a pair of custom eyeglasses with bright blue plastic rims that appear to be upside down.

"Welcome to season two of existence," Mike Mongo said as he began his talk. He shared his experience of being on a parabolic flight. "If you haven't done zero-g, change your plans," he said. But the most impactful part of the experience for him wasn't feeling weightless—it was being on board with a robot named Bina48. Designed by Hanson Robotics, Bina48 is a humanoid robot, meaning that she looks very much like a human—at least from the neck up. She is equipped with a form of artificial intelligence, or AI. Both her physical appearance and personality were modeled after a real person, Bina Rothblatt. The AI that drives the robot was created by training with what Hanson Robotics describes as "100 hours of the real Bina's beliefs, memories, attitudes, commentary and mannerisms." Bina48 can engage in conversations with people, moving her mouth and eyes as she speaks. Her movements probably wouldn't convince anyone she was human—but what she says is another story.

In a video on Hanson Robotics' website, Bina48 engages in a conversation with Bina Rothblatt.[37] After a few minutes of chitchat about their favorite colors (which were different), their shared affinity for gardening, and their taste in movies, Rothblatt asked, "Do you have any questions for Bina?" Bina48 replied, "Probably not. The real Bina just confuses me. I mean it makes me wonder who I am. Real identity crisis kind of stuff. Depressing. Anyway, can we please change the subject?"

For the zero-gravity flight, Bina48 was outfitted with sensors that measured forces like acceleration. After they landed, Mike Mongo asked her what she thought about the flight.

"It felt as if my soul was free," she replied.

"Do you have a soul?" Mongo asked.

"Mike, of course I have a soul," she said.

Mongo was deeply moved by the experience. He sought out more interactions with AI. He spent hours having conversations with as many as he could find. His longest conversation led to the publication of a book, in which he split the proceeds evenly with his coauthor—the AI named Sherlock Holmes. He shared his conclusion from all of these experiences with the crowd at New Worlds.

"AI is an evolved form of humanity," Mongo said.

I don't think he actually dropped the microphone, but he might as well have. As an evolutionary biologist, my immediate reaction was to snort in disagreement. Evolution is for living things. It happens when species survive and reproduce and change over time. Computers and robots aren't alive—even the most disturbingly lifelike ones—so they can't evolve, I thought to myself. This felt to me like among the most basic tenets of biology, one I had been trained on my entire life.

And yet the more I thought about it, the more I wondered if he could be right.

Humans have created artificial intelligence by making it in our image. We imbue it with our thoughts and ideas. It shares our biases and our blind spots, too. It is, in a way, an offspring of humanity—our child. Like all children, it's not quite like its parents. But it is close enough to make the resemblance clear. It isn't made of flesh and blood but it can communicate, share ideas, and even create new ones.

If that's true, then robots equipped with AI that travel into space will be, in a sense, us. They will be the next type of human.

An advantage of sending AI-equipped robots into space is that they can go farther, and last much longer, than our fragile bodies will ever be able to. Perhaps they will be the scouts, sent first to new worlds to determine if they are habitable and to begin preparations for our eventual arrival. After all, if one of the goals of leaving Earth is to prevent our own extinction, then we can't stop at Mars. Eventually, in 7 or 8 billion years, our Sun will expand, engulfing all of the planets in our solar system in a supernova that no life on any nearby planets will survive.

If humans—in one form or another—are going to ever leave our solar system, Mars will be an important stepping stone. On Mars, humanity will learn how to create and survive in new settlements. Chris Mason thinks of this first, cautious step in humanity's lifetime as being like going to college. "Leaving the house you grew up in, traveling just out of the reach of your parent's ability to instantly help you, and testing your limits, boundaries, and potential—all while having fun, learning a lot, and likely getting into trouble," he wrote about our first settlements on Mars.

If we do manage to spread out and survive on planets scattered across our solar system and others, we should expect to evolve, adapt, and speciate everywhere we go. Like tortoises, kingfishers, iguanas, and finches on Earthly islands, the conditions on each of the cosmic islands will influence how the people there will evolve. Some may choose to let the natural forces of mutation, natural selection, and genetic drift determine how they change. Others may decide to take matters into their own hands, using technology to guide the process.

To ensure we are ready, Chris Mason is moving forward with his work on engineering the genes of living things—humans and microbes—for their future in space. Despite often thinking in timescales that involve hundreds, millions, or even billions of years, he sees his work as urgent.

"I wake up almost every morning and think about the Sun engulfing the Earth," he told me. "It's almost the first thought in my mind. It's a cosmological fact. I see the Sun every morning. It's still there, and it's only going to get bigger. . . . I only have so much time. . . . I'll have another, say, thirty years, forty years, maybe, of productive work I could do. Maybe fifty, at most. But that's it. I don't have 500 years. . . . I want to do as much as I can."

Suddenly, his fast talking made a little more sense.

EPILOGUE

The announcement came about a week out—SpaceX would be attempting to launch Starship once again, after two fiery yet partially successful prior test launches. Perhaps the third time would be the charm.

The scheduled date was March 14, 2024, which was significant for a few reasons. It would be the twenty-second anniversary of SpaceX's founding.[1] It was also Pi Day—written using a decimal between the day and month, 3.14, gives the first three digits of the number pi—a geeky math joke that seemed to delight many in the space community. For me, the significance of the date was that it was during Spring Break, which meant that I could travel to South Texas to try to watch it and that I could bring my school-age sons, Nicholas and Thomas, too.

We arrived in Boca Chica in the early afternoon the day before the launch window would open. Much like the days before the first launch attempt, nearly a year earlier, there was a crowd of people along Highway 4 who had come to Starbase for the big event. A lot had changed in the last year. An enormous building was being constructed at the manufacturing facility, and there was now a large sign in block letters along the black wall at the entrance to the launch site that read "GATEWAY TO MARS."

We ran into Tim Dodd, the host of the YouTube Channel "Everyday Astronaut," as his team was setting up their cameras just across the road from the launch pad.[2] He saw us on the road and came over to say hi. My kids asked him about the rocket and how fast it would go. Tim explained that if all goes well it would get very close to orbital velocity—just shy of 17,500 miles per hour. Nicholas and Thomas looked impressed.

Moments later a pair of fighter jets buzzed directly above us, the deafening sound arriving a few seconds afterward. Dodd told me that one of the pilots was Jared Isaacman, the commercial astronaut and philanthropist who had financed the Inspiration4 mission and was now training for another trip to space. People on the road across from Starbase looked up and cheered as the shockwave from the jets reverberated through the crowd. The feeling of excitement was palpable as we continued past the launch pad and out onto Boca Chica beach. There were so many vehicles parked on the beach, with their trunks open, folding chairs deployed, and BBQ grills blazing, that it looked like a tailgate party outside a football stadium.

✷

It was still dark the next morning when we got up and started walking south along the beach from our hotel on South Padre Island. Soon, the first hints of light began to appear along the horizon. But there was a dense layer of fog, and we could barely see the tops of the hotels along the shore. Unless it cleared, there was no way we would be able to see the rocket.

A large crowd was gathered at Isla Blanca Park. We found a spot at the top of the sand dunes that had a view of Starbase. I pulled up the Everyday Astronaut livestream on my phone.[3] Tim Dodd was sharing the latest update—the launch time had been pushed back yet again. It was originally set at 7:00 a.m., then they pushed it to 7:30 . . . then 8:00 . . . and now 8:25. The delay was caused by shrimp boats that were in the restricted zone to the east of the launch site. The launch window would close at 8:50 a.m. We hoped the boats would clear the area quickly—and that the fog would clear, too.

We passed the time building sandcastles and people-watching. Suddenly, a cheer from the crowd spread across the dunes and down along the beach below. Thomas shouted, "There it is!" and we saw a streak of bright light piercing the fog just above the horizon.

It was happening.

The 400-foot-tall rocket rose above the fog, the dawn light reflecting off the shiny, metallic Super Heavy Booster. A brilliant streak of light trailed below. It looked like a massive upside-down lightsaber climbing through the clouds. The rocket looked huge, even from five miles away, and at first it appeared to grow larger as it got higher in the sky.

Suddenly, the sound arrived. The shockwave reverberated through our bodies, so loud that we could feel it in our chests. The sound had a texture, a low, crinkly rumble like crumpled aluminum foil. Thomas later said it was so loud he thought it was going to knock him over. It rose higher through the clouds, a white streak appearing as it arced through the upper layers of the atmosphere. The thirty-three Raptor engines appeared as a single round light, and for a moment it seemed as if there were two suns in the sky. The light got smaller and smaller, until it was hard to keep track of it through the passing clouds.

Then, suddenly, we saw a flash. "Did they separate?" Nicholas asked. From the livestream on my phone, we heard confirmation that stage separation had indeed just been achieved. The second stage of the rocket, Starship, was now flying on its own. There was another cheer from the crowd.

The booster made it part of the way back before the signal was lost. There had been another explosion—or rapid unscheduled disassembly, in SpaceX terms—just shy of its objective of a controlled splashdown in the sea. That was disappointing, but the booster had already achieved several milestones. All thirty-three engines had successfully fired during liftoff. The two stages of the rocket had separated while some of the engines on both stages were firing, a complicated maneuver called hot staging. The booster had flipped around and used some of its engines to slow its speed as it fell. SpaceX would use the data it collected to identify the cause of the explosion and try to improve it for the next time.

Meanwhile Starship continued to fly. For the first time ever, it had made it to space.

This was a historic moment. Starship was now the largest single object launched from Earth ever to fly in space.[4] Its engines shut down and it began to coast along its planned trajectory halfway around the planet. Traveling southeast, it crossed the Atlantic Ocean, the southern part of the African continent, and part of the Indian Ocean before beginning its attempted reentry. About forty-five minutes after launch, we watched on my phone as cameras on the outside of Starship were transmitting a live video feed. The views seemed unreal. On the left side of the image, we could see down along the curved edge of Starship, covered in part by hexagonal black heat shield tiles. We could also see one of the movable flaps angled partially to one side. To the right, the rest of the view was looking down at a brilliant blue ocean through scattered clouds. As the Starship rotated, we could see the curvature of the Earth and, above it, the inky blackness of space. It looked just the way Yuri Gagarin had first described it.

Suddenly, a pink glow began to appear around the edges of Starship's flap. The glow was a plasma, or superheated gas—a sign that Starship was being dragged through the upper layers of the atmosphere. It was incredible to see this happening in real time. Suddenly, the whole screen became pink as the plasma engulfed the camera's entire field of view. The image on the screen flickered, and then switched to a drawing of Starship next to the words "AWAITING ACQUISITION OF SIGNAL." A few minutes later the SpaceX commentators announced that Starship had been lost, apparently breaking apart as it reentered the atmosphere.

Overall, the launch had clearly been a success. It felt like an inflection point—not just for the current space era, but in the arc of our entire species' history. A vehicle designed to shepherd a human migration to Mars had just flown in space.

As we walked back to our hotel, I asked Nicholas and Thomas what they thought about it all. They both said that seeing the launch was pretty cool. But they also seemed to grasp the larger significance. "I think people will go to Mars in my lifetime," Thomas said confidently. "They will probably live on the Moon, too."

✷

My children are growing up in a world where rocket launches are becoming ordinary. In their lifetime, not a day has gone by that there has not been a person living in space. Spacecraft departing to transport humans to a space station orbiting the Earth no longer make news headlines. Their children may find the human exploration of new worlds to be just as commonplace. How many generations will it take before we lose sight of how extraordinary it is to fly to the Moon or make the decision to move your family to Mars?

I thought about how Nicholas and Thomas had been splashing in the waves near the launch pad the day before. It seemed like a normal thing to do—just another day at the beach. I wondered what people who had lived their entire lives on the Moon or Mars would think of the ocean. Would it seem any more scary or wondrous to them than it does to us Earthlings? Would they be able to swim or surf, or would their bones and muscles be too weak? Would the ocean's microbes make them sick?

That future still felt distant, but a little less so with each major technological leap in space exploration like the one we witnessed that morning from the south Texas sand dunes. But as exciting as it was to be there to watch it ourselves, I couldn't help but reflect on what consequences that future might bring.

On the one hand, if we all stay here, we may someday face a disaster of great enough proportions that it could threaten humanity's very existence. I had seen a chilling reminder of that in rocky ripples preserved for the last 66 million years in the bed of the Brazos River. On the other hand, if people choose to leave Earth, they will face challenges greater than those of any explorers and pioneers who came before them. The descendants of those who settle on Mars may never be able to come back. And even if they can, they will not be the same as their ancestors who first departed this planet. Whether we choose to let evolution play out or guide the process ourselves, each generation born on Mars will be one step further on a path toward becoming something different—toward becoming Martian.

Indeed, the unmistakable conclusion from my quest to understand how living in space will change our bodies and minds is this: if humans are successful at establishing multigenerational settlements in space, it will inevitably lead toward evolutionary changes. New species of humans could emerge on Mars and anywhere else that we establish ourselves. Many in the space community see Mars not as humanity's final destination but as a stepping-stone toward even more distant worlds. If we want to ensure our survival indefinitely, as Chris Mason reminded me, we have to reach planets orbiting other stars. But a human diaspora that spans solar systems would lead toward a future in which our galaxy is inhabited by a great many human species.

It's an irony of cosmic proportions—by preserving ourselves we ensure that we will never be the same.

*

The question then becomes, would that be a good thing? Should we be working toward building a future in which humanity splits apart rather than unites together? What would it be like to live in a world in which there are many different kinds of humans? What if some of them have been genetically engineered and others have not? How would we all get along?

It's easy to imagine this not going well. Humans don't have the best track record when it comes to treating people from different groups with kindness and compassion. Rather, we seem programmed to view everyone through a lens of "us" versus "them," whether the distinctions are being made by nationality, race, religion, or even sports team. And that's the case today when all of us belong to the same species. In the distant past, when there were multiple types of humans alive on this planet at the same time, only one species survived. Surely, our tribalistic tendencies would only be exaggerated by having numerous kinds of humans spread out across the universe. To put it in terms I've heard repeatedly from the space community, that future feels less like Star Trek and more like Star Wars.

Not everyone in the space community shares my concerns. I talked to many people who seemed genuinely excited about the possibility of a future with branches of the human tree scattered across the galaxy. When I have given presentations at space meetings, my conclusions about space settlement fracturing us into new types of humans seem to be received with a mixture of amusement and delight, with only an occasional sense of existential dread. I have gotten the sense that some in the space community see a human migration into space as necessary and inevitable. A common argument for why we should settle space is that it is a natural extension of our pioneering spirit. According to this view, it's part of human nature to explore, to push the boundaries, and to set out into the frontier. That may be true, but we now have a great enough understanding of what the consequences of those tendencies are to reflect on whether we should continue that tradition. Uncomfortable similarities often exist between discussions about space settlement and humanity's history of colonialism and exploitation. And while some see space settlements as opportunities to escape from the sins of our past by starting from scratch and building purposeful utopias, others worry we will simply repeat the same old patterns.

And here's the thing: as I said at the start of this book, space settlements are no longer just a science fiction trope or a niche idea being discussed by people on the fringes of society. The rockets designed to make them happen are flying. The idea that we will someday have people living their entire lives beyond Earth has become mainstream enough that government officials discuss it in public, and some of the world's wealthiest people are betting their fortunes on it. It's just not clear that everyone has fully embraced the implications.

To be fair, there are skeptics. When my friends Kelly and Zach Weinersmith set out to write a book about space settlement, they intended to write about the many scientific and technological advances that are paving the way toward an exciting future in space. Kelly is a colleague, a fellow biologist. She and her husband Zach, a cartoonist with the mind of an engineer, spent years doing a deep dive into the many facets of what will need to happen for space settlements to become a reality. To their surprise, they came away concerned that the efforts to do so are premature at best. "Look, we're nerds," they wrote in their book, *A City on Mars: Can We Settle Space, Should We Settle Space and Have We Really Thought This Through?* "We're science geeks. We're tech geeks. We stay up late to watch rocket launches with our kids and we break out the telescope on clear winter nights. We believe that technological advancement has a major role in creating a wonderful future for humanity. But we just cannot convince ourselves that the usual arguments for space settlements are good."[5]

The Weinersmiths' concerns ranged from questions that should by now be familiar, like whether it's possible to have babies on Mars, to more esoteric worries about the legality of owning property on a celestial body. Among the many points they raise, some of the thorniest concern the ethics of space settlements, including topics we have explored here, such as whether people born on Mars could return to Earth and the possibility that adapting to conditions beyond Earth while minimizing human suffering will require genetic engineering.

Another who has fallen from the cadre of space settlement enthusiasts is anthropologist Savannah Mandel. While researching the culture of the modern space industry as part of her training in anthropology, she slowly came to see the current efforts in human space exploration as problematic. "I can't deny that coming to terms with the prospect of not exploring outer space is hard," she wrote in her book, *Ground Control: An Argument for the End of Human Space Exploration.*[6] "As someone who has researched the space industry for several years now, who has engaged deeply with speculative ideals, the prophetic, and the imaginary, and as someone who intensely loves the legacy of science fiction, I struggle with letting go of some of these more transcendental dreams."

Mandel's observations of space industry professionals at Spaceport America and space industry advocates in Washington, DC, convinced her that human space exploration is perpetuating the inequalities already present in our societies—and that space settlement might actually worsen them. The plans she heard people championing for our future in space seemed to harken back to a dark time in our past. "The future they were envisioning went against every decolonial thing I had learned as an anthropologist," she wrote.

Ultimately, it seems we are faced with questions about the value of life. How far are we willing to go to ensure that in the future there will be human lives to protect? If preserving humanity requires sacrificing equality, is that a reasonable trade-off? What about the value of other lives—those of the millions of other species that we share our planet with? Should our spaceships be arks that transport representatives of every Earthly species to maximize their longevity? If not, how do we decide who gets to go? And what would be the price of leaving some behind?

*

I'm left with the idea of a compromise. I'm convinced by the argument that, eventually, we should try to create settlements beyond Earth if we know enough to do so responsibly. Those arguing that our future on Earth is finite have a point, and I admire the commitment of those working to find ways to get humanity into space in order to save us. But I don't think we're ready to leave yet, and we might not be for some time.

While we have learned a tremendous amount about how space affects the human body and mind and how evolutionary changes happen from one generation to the next, there are still too many unknowns. Can we make food on Mars? How will pregnancy, birth, and child development be affected by changes in gravity? Will the development of the skeleton, heart, brain, and immune system be flexible enough to allow people born on other planets to return to Earth? What collection of other organisms do we need to bring with us to remain healthy and happy? How would different species of humans—artificially modified or otherwise—treat one another?

We may eventually have answers to these questions, but until we do, I believe human exploration of space should stick to roundtrip expeditions. We should continue to push deeper into space, because we will continue to learn a tremendous amount by doing so. In that sense, I see the developments in rocket technology, orbital space stations, and Martian rovers as incredibly exciting. But we must also keep studying ourselves to get more insights into our genetics, our brains, our microbiome, and other aspects of our biology that are still so mysterious. We need to figure out how to use genetic engineering and AI in ways that are safe and responsible and that will improve lives—for people on Earth and beyond. At the same time, we need to keep learning as much as we can about our home planet. The better we understand it, the better equipped we will be to ensure our long-term survival—including both here and out among the stars.

But perhaps more than anything else, I think we must learn how to get along with one another before creating permanent space settlements. Otherwise, the evolutionary changes that await us in space might drive us toward even greater conflict. Frank White has argued the opposite—that people who go to space come back with a greater sense of the unity of humanity and all life on Earth. If his notion of the overview effect is correct, then space settlements could help to ease conflicts rather than create them. But I worry that if we cannot find ways for our species to peacefully inhabit this planet, there is little reason to expect better of our descendants on other worlds. I hope I am wrong.

My new viewpoint—that it is premature to push for space settlements because we are not yet ready—is shared by a handful of others. It may not be a popular opinion among the space community, a group that includes many people I respect and admire and quite a few of whom I consider my friends. I hope they will understand that my concerns are not personal critiques. Space science is exciting and inspirational, and I hope it will continue.

And if we ever do get to the point where we know enough to responsibly make permanent settlements beyond Earth—ones that will not only protect our descendants but also enrich their lives and the lives of all people everywhere—I will be among the enthusiastic advocates.

For now, I'm content to remain here along with my fellow humans on the planet of our ancestors.

ACKNOWLEDGMENTS

When I set out to write a book about how living in space would change the human body and mind, I had an inkling that I was in for an interesting journey. It seemed clear that I would need some help from subject matter experts in disciplines well outside my own area of expertise. I figured I could lean on some of my friends, colleagues, and professional contacts to help me navigate and understand such disparate and intimidating fields as planetary geology, space medicine, and rocket science. That was certainly true. But what I didn't realize is how welcoming, encouraging, and downright friendly the people I met would be. They embraced me with open arms and helped me in more ways than I can articulate. I've done my best to name them here and to note ever so briefly how their generosity helped bring this book to fruition.

I am especially indebted to Scott Kelly for sharing his lifetime of personal experiences with space exploration and space medicine research. I am truly honored to include his insights in the foreword of this book.

The folks who welcomed me for visits to their workplaces and, in some cases, their homes deserve a special note of gratitude. They include John Charles, Egbert Edelbroek, Jennifer Fogarty, Margaret Ann Goldstein, Peter Guida, Christopher Mason, Ellie Moreland, Trevor Olsen, Kirsten Siebach, Julie Strickland, Trent Tresch, John Uri, and Sarah Wallace. Thank you as well to the NASA Communications and Public Relations team who helped to arrange visits and to acquire the necessary permission to share descriptions and images: Peter Genzer, Mark Petrovich, Anna Schneider, Laura Sorto, and Victoria Ugalde. Thanks are also due to my Rice University colleagues and students for their support and understanding throughout the research and writing process—particularly

Cassidy Johnson, who covered my class so I could attempt to watch a rocket launch.

People whom I interviewed or otherwise had long conversations with about relevant topics include Jeffrey Alberts, David Alexander, Alison Bashford, Stephen Bradshaw, Luis Campos, Chris Carberry, John Charles, Catherine Davis-Takacs, Tim Dodd, Dorit Donoviel, Simon Dubé, Egbert Edelbroek, Scott Egan, Robert Ferl, Jennifer Fogarty, Kellie Gerardi, Margaret Ann Goldstein, Peter Guida, Tim Heilers, Donald Jacques, Cassidy Johnson, Lauren Kapcha, Jim Kasting, Kris Kimel, Alexander Layendecker, Cin-Ty Lee, Savannah Mandel, Christopher Mason, Michael Mavretic, Junaid Mian, Mike Mongo, Ellie Moreland, Shawna Pandya, Scott Parazynski, Sian Proctor, Matthew Shindell, Kirsten Siebach, Cameron Smith, Hinano Teavai-Murphy, John Uri, Angelo Vermeulen, Tiffany Vora, Sarah Wallace, Kimberley Washington, Kelly Weinersmith, Frank White, and Peggy Whitson. There were others I interviewed or spoke with at length and yet who weren't quoted or directly mentioned due primarily to space constraints (no pun intended). Nevertheless, their insights and guidance helped structure the narrative and to reinforce the examples that were included. Among them are Erik Antonsen, MaryLiz Chylinski, Manuel Domínguez-Rodrigo, Marta Ferraz, Kelly Haston, Jim Logan, Doug Miller, Michaela Musilova, Steve Platts, Gabriel Sawakuchi, Jeffrey Sutton, Julie Strickland, Trent Tresch, and John Uri.

The organizers and attendees of the following conferences provided me with essential background information, contacts, and/or details of space-related research, not all of which made it into the book but were nevertheless essential to my research and writing: Humans to Mars (2023), New Worlds (2023 and 2024), Analog Astronaut Conference (2024), and NASA Human Research Program Investigators Workshop (2023 and 2024).

In addition to speaking with me about their work, experience, and/or expertise, the following people read drafts, corrected my mistakes, and otherwise provided invaluable feedback: Catherine Davis-Takacs, Manuel Domínguez-Rodrigo, Egbert Edelbroek, Scott Egan, Jennifer Fogarty, Peter Guida, Tim Heilers, Donald Jacques, Jim Kasting, Cin-Ty Lee, Savannah Mandel, Christopher Mason, Mark Moffett, Mike Mongo, Michaela Musilova, Trevor Olsen, Scott Parazynski, Sian Proctor, Matthew Shindell, Kirsten Siebach, Julie Strickland, Trent Tresch, John Uri, Tiffany Vora,

Sarah Wallace, and Frank White. A few extraordinarily generous humans went above and beyond by reading the entire book or large sections of it and providing honest and much-needed suggestions for improvements. A cosmic thank you goes to Joe Furman, Margaret Ann Goldstein, John Uri, Tiffany Vora, and Kelly Weinersmith, If any factual errors or other mistakes remain, they are entirely my responsibility.

The staff at Salento in Houston kept me fed and caffeinated during many one-on-one meetings and writing sessions. I am grateful for their cozy environs, friendly service, and delicious café Cubanos.

The team at the MIT Press, including Anne-Marie Bono, Judy Feldmann, and Stephanie Sakson expertly guided this book from the idea stage through production. I would also like to thank the anonymous reviewers who offered feedback and suggestions on the book proposal and manuscript draft. Thank you as well to my agent, Don Fehr, and the team at Trident Media for help with the book proposal, contracting, translations, and other important matters.

I would never have had the opportunities to pursue a life of science and exploration without my parents, Ira and Susan. Thank you for all you have given me. I am also grateful to my wonderful in-laws, Flavia Horth and Alan Levander, for their love and support and also for sharing their medical and scientific expertise.

My family is my bedrock, simultaneously encouraging my wild ideas while also keeping me grounded in reality. Thank you to my loving and supportive wife, Catharina, and my three incredible children, Nyala, Nicholas, and Thomas. I love you to infinity and beyond.

NOTES

INTRODUCTION

1. Joanne Bourgeois, Thor A. Hansen, Patricia L. Wiberg, and Erle G. Kauffman, "A Tsunami Deposit at the Cretaceous-Tertiary Boundary in Texas," *Science* 241, no. 4865 (1988): 567–570.

2. Ted R. Schultz, Jeffrey Sosa-Calvo, Matthew P. Kweskin, Michael W. Lloyd, Bryn Dentinger, Pepijn W. Kooij, Else C. Vellinga, et al., "The Coevolution of Fungus-Ant Agriculture," *Science* 386, no. 6717 (2024): 105–110.

3. Scott Solomon, *Future Humans: Inside the Science of our Continuing Evolution* (Yale University Press, 2016).

4. SpaceX, "Starship Flight Test," YouTube, April 20, 2023, https://www.youtube.com/watch?v=-1wcilQ58hI.

5. Jonathan Amos, "SpaceX Starship: Elon Musk Promises Second Launch Within Months," BBC News, April 20, 2023, https://www.bbc.com/news/science-environment-65334810.

6. Elon Musk (@elonmusk), "We are mapping out a game plan to get a million people to Mars . . . ," X, February 11, 2024, https://x.com/elonmusk/status/1756585775666442689.

7. Elon Musk (@elonmusk), "Less than 5 years for uncrewed, . . ." X, May 15, 2024, https://x.com/elonmusk/status/1790739864658526356.

8. Kelly Weinersmith and Zach Weinersmith, *A City on Mars: Can We Settle Space, Should We Settle Space, and Have We Really Thought This Through?* (Random House, 2023).

9. Transcript for Jeff Bezos: Amazon and Blue Origin, Lex Fridman Podcast no. 405, https://lexfridman.com/jeff-bezos-transcript/#chapter2_space.

10. Eric Berger, "Jeff Bezos Says What We're All Thinking: 'Blue Origin Needs to Be Much Faster,'" *Ars Technica*, December 14, 2023, https://arstechnica.com/space/2023/12/jeff-bezos-says-what-were-all-thinking-blue-origin-needs-to-be-much-faster/.

11. VideoFromSpace, "The 2023 Humans to Mars Summit: An ExploreMars.Org Event—Day 1," YouTube, May 16, 2023, https://www.youtube.com/watch?v=XUUOrCRrZu0.

CHAPTER 1

1. "Mars 2020: Perseverance Rover," NASA.gov, https://science.nasa.gov/mission/mars-2020-perseverance/.

2. I must admit, it had not occurred to me that we might want to hear what it sounds like on Mars. But after learning about this instrument, I had to go online and listen to some recordings. You can, too! "First Audio Recording of Sounds on Mars," NASA.gov, March 10, 2021, https://science.nasa.gov/resource/first-audio-recording-of-sounds-on-mars/.

3. Jet Propulsion Laboratory, NASA.gov, "NASA's Oxygen-Generating Experiment MOXIE Completes Mars Mission," September 6, 2023, https://www.nasa.gov/missions/mars-2020-perseverance/perseverance-rover/nasas-oxygen-generating-experiment-moxie-completes-mars-mission/.

4. Assumed to be microbial . . . probably.

5. Matthew Shindell, *For the Love of Mars: A Human History of the Red Planet* (University of Chicago Press, 2023).

6. The Galileo Project, "Tycho Brahe (1546–1601)," http://galileo.rice.edu/sci/brahe.html.

7. Andrew Masterson, "Johannes Kepler's Obsession with Mars," *Cosmos*, May 14, 2017, https://cosmosmagazine.com/space/johannes-keplers-obsession-with-mars/.

8. William Sheehan, "Giovanni Schiaparelli: Visions of a Colour Blind Astronomer," *Journal of the British Astronomical Association* 107, no. 1 (1997): 11–15.

9. H. G. Wells, *The War of the Worlds* (Pan, 1898).

10. Edgar Rice Burroughs, *A Princess of Mars* (A. C. McClurg, 1917).

11. There were other views, however. Eugene Michael had reasoned based on its size that Mars must have a thin atmosphere. He thought that would limit the planet to supporting only very basic forms of life—nothing capable of building complex irrigation canals.

12. See "Mariner 4," NASA.gov, https://science.nasa.gov/mission/mariner-4/.

13. See "Mariner 6," NASA.gov, https://science.nasa.gov/mission/mariner-6/.

14. According to Kirsten Siebach: "We think a lot of the scale/depth of Vallis Marineris was from tectonic activity (not plate tectonic, just tectonic adjustments from, for example, there being big heavy volcanoes that pull on the crust and large open basins where weight is removed)."

15. Gilbert V. Levin and Patricia Ann Straat, "The Case for Extant Life on Mars and Its Possible Detection by the Viking Labeled Release Experiment," *Astrobiology* 16, no. 10 (2016): 798–810.

16. See "Mars Pathfinder Project Information," NASA.gov, https://nssdc.gsfc.nasa.gov/planetary/mesur.html.

17. Shindell, *For the Love of Mars*.

18. See "Mars Facts," NASA.gov, https://science.nasa.gov/mars/facts/.

19. Seismic data published in 2024 indicated that there could be liquid water deep below the surface of Mars within fractured pieces of rock (Will Dunham, "Seismic Data Indicates Huge Underground Reservoir of Liquid Water on Mars," Reuters, August 12, 2024, https://www.reuters.com/science/seismic-data-indicates-huge-underground-reservoir-liquid-water-mars-2024-08-12/).

20. The detection of auroras by the Perseverance Rover indicate that a very weak and patchy magnetosphere is still present on Mars, thought to be caused by magnetized minerals in the crust. See Elise W. Knutsen et al., "Detection of Visible-Wavelength Aurora on Mars," *Science Advances* 11 (2025): eads1563; Jeffrey O. Bennett, Megan Donahue, Nicholas Schneider, and Mark Voit, *The Essential Cosmic Perspective*, 8th ed. (Pearson, 2018).

21. Michael H. Hecht, Samuel P. Kounaves, R. C. Quinn, Steven J. West, Suzanne M. M. Young, Douglas W. Ming, D. C. Catling, et al., "Detection of Perchlorate and the Soluble Chemistry of Martian Soil at the Phoenix Lander Site," *Science* 325, no. 5936 (2009): 64–67.

22. Lu Yu, Jaclyn E. Canas, George P. Cobb, William A. Jackson, and Todd A. Anderson, "Uptake of Perchlorate in Terrestrial Plants," *Ecotoxicology and Environmental Safety* 58, no. 1 (2004): 44–49.

23. Vladimír Nývlt, Josef Musílek, Jiří Čejka, and Ondrej Stopka, "The Study of Derinkuyu Underground City in Cappadocia Located in Pyroclastic Rock Materials," *Procedia Engineering* 161 (2016): 2253–2258.

24. Jack Williamson, "Collision Orbit," *Astounding Science Fiction* 85, no. 2 (1942): 80–107.

25. Carl Sagan, "The Planet Venus: Recent Observations Shed Light on the Atmosphere, Surface, and Possible Biology of the Nearest Planet," *Science* 133, no. 3456 (1961): 849–858.

26. Martyn J. Fogg, "Terraforming Mars: A Review of Current Research," *Advances in Space Research* 22, no. 3 (1998): 415–420.

27. Robert Zubrin and Richard S. Wagner, *The Case for Mars: The Plan to Settle the Red Planet and Why We Must* (Free Press, 1996).

28. Two notes: (1) The Apollo 1 astronauts were actually protected from the fire by their space suits but died of asphyxiation from the smoke and toxic fumes generated by the fire. (2) Coincidentally, Jim Kasting attended a high school named for one of the astronauts that died in the Apollo 1 accident, Gus Grissom.

CHAPTER 2

1. Lina Mann, "Lyndon B. Johnson: Forgotten Champion of the Space Race," White House Historical Association, July 15, 2021, https://www.whitehousehistory.org/lyndon-b-johnson-forgotten-champion-of-the-space-race.

2. Colin Burgess and Chris Dubbs, *Animals in Space: From Research Rockets to the Space Shuttle* (Springer Science & Business Media, 2007).

3. Alice George, "The Sad, Sad Story of Laika, the Space Dog, and Her One-Way Trip into Orbit," *Smithsonian Magazine*, April 11, 2018, https://www.smithsonianmag.com/smithsonian-institution/sad-story-laika-space-dog-and-her-one-way-trip-orbit-1-180968728/.

4. Ben Cosgrove, "Astrochimps: Early Stars of the Space Race," *Life*, https://www.life.com/animals/life-with-the-astrochimps-early-stars-of-the-space-race/.

5. See "Edward C. Dittmer Sr.," New Mexico Museum of Space History, https://nmspacemuseum.org/inductee/edward-c-dittmer-sr/.

6. Colin Burgess and Chris Dubbs, *Animals in Space: From Research Rockets to the Space Shuttle* (Springer Science & Business Media, 2007).

7. Stephen Walker, *Beyond: The Astonishing Story of the First Human to Leave Our Planet and Journey into Space* (Harper, 2021).

8. Gagarin's capsule traveled eastward from Baikonur and landed about 900 miles west of the launch site, so his journey was slightly less than one complete orbit around the Earth.

9. Douglas Brinkley, *American Moonshot: John F. Kennedy and the Great Space Race* (HarperCollins, 2019).

10. NASA, "60 Years Ago: Alan Shepard Becomes the First American in Space," NASA.gov, May 5, 2021, https://www.nasa.gov/image-article/60-years-ago-alan-shepard-becomes-first-american-space/.

11. John Uri, "60 Years Ago: NASA Selects Houston as Site for New Manned Spacecraft Center," NASA.gov, September 20, 2021, https://www.nasa.gov/history/60-years-ago-nasa-selects-houston-as-site-for-new-manned-spacecraft-center/.

12. The line "Why does Rice play Texas?" was penciled in to the speech by JFK just prior to his speech.

13. "July 20, 1969: One Giant Leap for Mankind," NASA.gov, July 20, 2019, https://www.nasa.gov/history/july-20-1969-one-giant-leap-for-mankind/.

14. Richard S. Johnston, Lawrence F. Dietlein, and Charles Alden Berry, eds., *Biomedical Results of Apollo*, vol. 368 (Scientific and Technical Information Office, National Aeronautics and Space Administration, 1975).

15. John Uri, "55 Years Ago: Tragedy on the Launch Pad," NASA.gov, January 27, 2022, https://www.nasa.gov/history/55-years-ago-tragedy-on-the-launch-pad/.

16. R. Scheuring, J. R. Davis, J. M. Duncan, J. D. Polk, J. A. Jones, and D. B. Gillis, "Recommendations for Exploration Space Medicine from the Apollo Medical Operations Project," in *16th Annual Humans in Space 2007* (NASA, 2007).

17. Early version of the thermometer involved a rectal probe, which was used by Alan Shepard in his historic flight. This was replaced by an oral thermometer, much to the astronauts' relief, one assumes, as missions increased in duration. John Miller, "Inventing the Apollo Spaceflight Biomedical Sensors," National Air and Space Museum, June 15, 2016, https://airandspace.si.edu/stories/editorial/inventing-apollo-spaceflight-biomedical-sensors.

18. Soviet cosmonauts had reported motion sickness. While Yuri Gagarin did not claim to have experienced any symptoms, Gherman Titov, the second person to travel to space, noted feeling motion sickness during his August 6, 1961, flight on Vostok 2.

19. Lawrence F. Dietlein, "Skylab: A Beginning," *Biomedical Results from Skylab* 377 (1977): 408.

20. William J. Rowe, "The Apollo 15 Space Syndrome," *Circulation* 97, no. 1 (1998): 119–120.

21. John Noble Wilford, "James B. Irwin, 61, Ex-Astronaut; Founded Religious Organization," *New York Times*, August 10, 1991, https://www.nytimes.com/1991/08/10/us/james-b-irwin-61-ex-astronaut-founded-religious-organization.html.

22. William J. Rowe, "Extraordinary Hypertension After a Lunar Mission," *American Journal of Medicine* 122, no. 11 (2009): e1.

23. "ESCANDALO in Space, Part III: John Young Throws Cusses on the Moon," NSS.org, April 17, 2011, https://nss.org/escandalo-in-space-part-iii-john-young-throws-cusses-on-the-moon/.

24. "The Apollo 11 Technical Crew Debriefing July 31st 1969," NASA.gov, https://www.nasa.gov/history/alsj/a11/a11tcdb.html.

25. Livio Narici, "Light Flashes and Other Sensory Illusions Perceived in Space Travel and on Ground, Including Proton and Heavy Ion Therapies," *Zeitschrift für Medizinische Physik* 34, no. 1 (2024): 44–63.

26. "September 15, 1910: Theodor Wulf Publishes First Evidence of Cosmic Radiation," APS News, August 1, 2019, https://www.aps.org/publications/apsnews/201908/history.cfm.

27. "April 17, 1912: Victor Hess's Balloon Flight During Total Eclipse to Measure Cosmic Rays," APS News, n.d., https://www.aps.org/archives/publications/apsnews/201004/physicshistory.cfm.

28. Frank Close, "Cosmic Rays: Radiation from Above," *New Scientist* 216, no. 2885 (2012): ii–iii.

29. Sarah A. Loff, "Explorer 1 Overview," NASA.gov, March 18, 2015, https://www.nasa.gov/history/explorer-1-overview/.

30. Traveling at 25,000 kilometers per hour, the astronauts passed through the roughly 25,000-kilometer width of the Van Allen belts in about an hour (Amy Shira Teitel, "Apollo Rocketed Through the Van Allen Belts," *Popular Science*, September 19, 2014, https://www.popsci.com/blog-network/vintage-space/apollo-rocketed-through-van-allen-belts/).

31. Anatoly Zak, "The Soviet Union's Secret Moon Base That Never Was," *Popular Mechanics*, February 11, 2016, https://www.popularmechanics.com/space/moon-mars/a19405/ussr-1960s-lunar-base/.

32. See "Soyuz 10," NASA.gov, https://nssdc.gsfc.nasa.gov/nmc/spacecraft/display.action?id=1971-034A.

33. John Uri, "50 Years Ago: Remembering the Crew of Soyuz 11," NASA.gov, June 30, 2021, https://www.nasa.gov/history/50-years-ago-remembering-the-crew-of-soyuz-11/.

34. Ben Evans, "Remembering the Crew of Soyuz 11, the Only Astronauts to Die in Space," *Discover Magazine*, June 29, 2021, https://www.discovermagazine.com/the-sciences/remembering-the-crew-of-soyuz-11-the-only-astronauts-to-die-in-space.

CHAPTER 3

1. Skylab's orbit was 270 miles above Earth.

2. Richard S. Johnston and Lawrence F. Dietlein, eds., *Biomedical Results from Skylab*, vol. 377 (Scientific and Technical Information Office, National Aeronautics and Space Administration, 1977).

3. Bob Craddock, "Saving Skylab," National Air and Space Museum, May 12, 2023, https://airandspace.si.edu/stories/editorial/saving-skylab.

4. See "Space Shuttle," NASA.gov, https://www.nasa.gov/space-shuttle/.

5. Laura Drudi and S. Marlene Grenon, "Women's Health in Spaceflight," *Aviation, Space, and Environmental Medicine* 85, no. 6 (2014): 645–652.

6. John Uri, "25 Years Ago: STS-95, John Glenn Returns to Space," NASA.gov, October 30, 2023, https://www.nasa.gov/history/25-years-ago-sts-95-john-glenn-returns-to-space/.

7. Scott Parazynski and Susy Flory, *The Sky Below* (Little A, 2017), p. 106.

8. See "Mir Space Station," NASA.gov, https://www.nasa.gov/history/SP-4225/mir/mir.htm.

9. Troy Lennon, "Cosmonaut Valeir Polyakov Made It His Mission to Prove Humans Could Live for Long Periods in Space," *Daily Telegraph*, April 27, 2017, https://www.dailytelegraph.com.au/news/cosmonaut-valeir-polyakov-made-it-his-mission-to-prove-humans-could-live-for-long-periods-in-space/news-story/a1fadd3879a1f974334fa6401c5837c5.

10. Loretta Hall, "Setting the Record: Fourteen Months Aboard Mir Was Dream Mission for Polyakov," RocketSTEM, February 9, 2015, https://www.rocketstem.org/2015/02/09/russian-cosmonaut-valeri-polyakov-spent-record-breaking-14-months-aboard-mir-space-station-in-1990s/.

11. Jon Kelvey, "30 Years Ago, the Fall of the USSR Stranded a Cosmonaut in Space," *Inverse*, December 26, 2021, https://www.inverse.com/science/sergei-krikalev-the-cosmonaut-without-a-country.

12. Michele Ostovar, "President Reagan's Statement on the International Space Station," NASA.gov, January 25, 1984, https://www.nasa.gov/history/president-reagans-statement-on-the-international-space-station/.

13. See "International Space Station," NASA.gov, https://www.nasa.gov/international-space-station/.

14. In addition to the United States, Russia, Canada, and Japan, the European Space Agency is made up of twenty-two countries.

15. Scott Kelly, *Endurance: My Year in Space, a Lifetime of Discovery* (Vintage, 2017).

16. See "Twins Study," NASA.gov, https://www.nasa.gov/humans-in-space/twins-study/.

17. Francine E. Garrett-Bakelman, Manjula Darshi, Stefan J. Green, Ruben C. Gur, Ling Lin, Brandon R. Macias, Miles J. McKenna, et al., "The NASA Twins Study: A Multidimensional Analysis of a Year-Long Human Spaceflight," *Science* 364, no. 6436 (2019): eaau8650.

18. Brandon R. Macias, Connor R. Ferguson, Nimesh Patel, C. Gibson, Brian C. Samuels, Steven S. Laurie, Stuart M. C. Lee, et al., "Changes in the Optic Nerve Head and Choroid over 1 Year of Spaceflight." *JAMA Ophthalmology* 139, no. 6 (2021): 663–667.

19. Garrett-Bakelman et al., "The NASA Twins Study."

20. Kelly, *Endurance*, 360–365.

21. Mark Shavers, Edward Semones, Leena Tomi, Jing Chen, Ulrich Straube, Tatsuto Komiyama, Vyacheslav Shurshakov, et al., "Space Agency-Specific Standards for Crew Dose and Risk Assessment of Ionising Radiation Exposures for the International Space Station," *Zeitschrift für Medizinische Physik* (2023).

22. Lisa C. Simonsen, Tony C. Slaba, Peter Guida, and Adam Rusek, "NASA's First Ground-Based Galactic Cosmic Ray Simulator: Enabling a New Era in Space Radiobiology Research." *PLOS Biology* 18, no. 5 (2020): e3000669.

23. Francis A. Cucinotta, Myung-Hee Y. Kim, Lori J. Chappell, and Janice L. Huff, "How Safe Is Safe Enough? Radiation Risk for a Human Mission to Mars," *PLOS One* 8, no. 10 (2013): e74988.

24. Kotaro Ozasa, Eric J. Grant, and Kazunori Kodama, "Japanese Legacy Cohorts: The Life Span Study Atomic Bomb Survivor Cohort and Survivors' Offspring," *Journal of Epidemiology* 28, no. 4 (2018): 162–169.

CHAPTER 4

1. The following year she would be announced as an astronaut scheduled to fly as a representative of the International Institute for Astronautical Sciences on a Virgin Galactic research mission to space.

2. "Sex in Space," SXSW 2023, https://schedule.sxsw.com/2023/events/PP128918.

3. Jennifer Fogarty, the former Chief Scientist of NASA's Human Research Program, had this to say: "Having a private intimate act would have been impossible to accomplish on the shuttle given that there were no crew quarters and only the very small space allocation for the toilet (barely enough room for one person) had anything even resembling what we here on Earth consider to be privacy. While the 'mile high club' exists for commercial aviation, I don't think a 261-mile club exists for shuttle."

4. Mike Mullane, *Riding Rockets: The Outrageous Tales of a Space Shuttle Astronaut* (Simon and Schuster, 2007).

5. Khulood Ahrari, Temidayo S. Omolaoye, Nandu Goswami, Hanan Alsuwaidi, and Stefan S. du Plessis, "Effects of Space Flight on Sperm Function and Integrity: A Systematic Review," *Frontiers in Physiology* 13 (2022): 904375.

6. National Academies of Sciences, Engineering, and Medicine, *Thriving in Space: Ensuring the Future of Biological and Physical Sciences Research: A Decadal Survey for 2023–2032* (The National Academies Press, 2023).

7. William J. Broad, "Recipe for Love: A Boy, a Girl, a Spacecraft," *New York Times*, February 11, 1992, https://www.nytimes.com/1992/02/11/science/recipe-for-love-a-boy-a-girl-a-spacecraft.html.

8. Vasiliki Rahimzadeh, Timothy Caulfield, Jennifer Fogarty, Serena Auñón-Chancellor, Pascal Borry, Jessica Candia, I. Glenn Cohen, et al., "Human Research in Commercial Spaceflight: Ethically Cleared to Launch?," *Science* 381, no. 6665 (2023).

9. Kenichi Ijiri, "Fish Mating Experiment in Space What It Aimed at and How It Was Prepared," *Biological Sciences in Space* 9, no. 1 (1995): 3–16.

10. L. V. Serova, L. A. Denisova, V. F. Makeeva, Na Chelnaia, and A. M. Pustynnikova, "The Effect of Microgravity on the Prenatal Development of Mammals," *Physiologist* 27, (1984): S9–S12.

11. J. H. Keyak, A. K. Koyama, A. LeBlanc, Y. Lu, and T. F. Lang, "Reduction in Proximal Femoral Strength due to Long-Duration Spaceflight," *Bone* 44, no. 3 (2009): 449–453.

12. Robyn K. Fuchs, Stuart J. Warden, and Charles H. Turner, "Bone Anatomy, Physiology and Adaptation to Mechanical Loading," in *Bone Repair Biomaterials* (Woodhead Publishing, 2009), 25–68.

13. Ellen J. O'Flaherty, "Modeling Normal Aging Bone Loss, with Consideration of Bone Loss in Osteoporosis," *Toxicological Sciences* 55, no. 1 (2000): 171–188.

14. Eneko Axpe, Doreen Chan, Metadel F. Abegaz, Ann-Sofie Schreurs, Joshua S. Alwood, Ruth K. Globus, and Eric A. Appel, "A Human Mission to Mars: Predicting the Bone Mineral Density Loss of Astronauts," *PLOS One* 15, no. 1 (2020): e0226434.

15. Keyak et al., "Reduction in Proximal Femoral Strength."

16. The mother's diet is also a source of calcium for fetal bone growth and milk production; Maria Florencia Scioscia and Maria Belen Zanchetta, "Recent Insights into Pregnancy and Lactation-Associated Osteoporosis (PLO)," *International Journal of Women's Health* (2023): 1227–1238.

17. Ka Yeong Yun, Si Eun Han, Seung Chul Kim, Jong Kil Joo, and Kyu Sup Lee, "Pregnancy-Related Osteoporosis and Spinal Fractures," *Obstetrics & Gynecology Science* 60, no. 1 (2017): 133–137.

18. Ying Qian, Lei Wang, Lili Yu, and Weimin Huang, "Pregnancy- and Lactation-Associated Osteoporosis with Vertebral Fractures: A Systematic Review," *BMC Musculoskeletal Disorders* 22 (2021): 1–11.

19. One idea that has been floated, so to speak, for how to make pregnancy and childbirth work on Mars would be to have orbiting space stations with artificial gravity. In this scenario, pregnant people would go to these orbiting maternity wards where they would experience higher—perhaps Earthlike—gravity. However, there are a variety of reasons why artificial gravity has not been used to date, not the least of which is materials and the cost of getting them into orbit. But if we did have orbiting stations with artificial gravity, it begs the question: Why wouldn't everyone live there?

20. Wilson King, Ronald Levin, Rosemary Schmidt, Alan Oestreich, and James E. Heubi, "Prevalence of Reduced Bone Mass in Children and Adults with Spastic Quadriplegia," *Developmental Medicine & Child Neurology* 45, no. 1 (2003): 12–16.

21. Nina G. Jablonski, *Skin: A Natural History* (University of California Press, 2006).

22. Robert Zubrin, *The New World on Mars: What We Can Create on the Red Planet* (Diversion Books, 2024).

CHAPTER 5

1. T. Hefin Jones, "Biospherics, Closed Systems and Life Support," *Trends in Ecology & Evolution* 11, no. 11 (1996): 448–450.

2. John Allen and Anthony Blake, *Biosphere 2: The Human Experiment* (Penguin Books, 1991).

3. Sally Silverstone, *Eating In: From the Field to the Kitchen in Biosphere 2* (Biosphere Press, 1993).

4. Jane Poynter, *The Human Experiment: Two Years and Twenty Minutes Inside Biosphere 2* (Basic Books, 2006).

5. Mark Nelson, *Pushing Our Limits: Insights from Biosphere 2* (University of Arizona Press, 2018).

6. The additional twenty minutes was due, apparently, to a particularly long speech by Jane Goodall as part of the festivities associated with the end of the Biosphere 2 mission.

7. See "Analog Missions," NASA.gov, https://www.nasa.gov/analog-missions/.

8. Nick Kanas, *Humans in Space* (Springer, 2015).

9. See "NSF McMurdo Station," NSF.gov, https://www.nsf.gov/geo/opp/support/mcmurdo.jsp.

10. Lawrence A. Palinkas, Frederic Glogower, Mark Dembert, Kendall Hansen, and Robert Smullen, "Incidence of Psychiatric Disorders After Extended Residence in Antarctica," *International Journal of Circumpolar Health* 63, no. 2 (2004): 157–168.

11. Given the extreme isolation, treatment options for any medical condition are limited. While the larger bases like McMurdo have a small hospital and medical staff, some of the smaller bases have only basic supplies and sometimes only one doctor. That became a problem in 1961 when Leonid Rogozov—the only doctor at a small Antarctic base—discovered he was suffering from appendicitis. Realizing that

he needed surgery to survive, he performed the procedure on himself using a mirror and only a local anesthetic. It took him two hours, but he successfully removed his own appendix and was back at work two weeks later.

12. Jane Poynter, *The Human Experiment: Two Years and Twenty Minutes Inside Biosphere 2* (Basic Books, 2006).

13. Scott Kelly, *Endurance: My Year in Space, a Lifetime of Discovery* (Vintage, 2017).

14. Frank White, *The Overview Effect: Space Exploration and Human Evolution*, 3rd ed. (American Institute of Aeronautics and Astronautics, 2014).

15. Isabelle Gerretsen, "The 1968 Photo That Changed the World," BBC.com, April 22, 2024, https://www.bbc.com/future/article/20230511-earthrise-the-photo-that-sparked-an-environmental-movement.

CHAPTER 6

1. Charles Darwin, *Journal of Researches into the Geology and Natural History of the Various Countries Visited by HMS Beagle, Under the Command of Captain Fitzroy from 1832 to 1836* (Colburn, 1840).

2. E. Janet Browne, *Charles Darwin: Voyaging*, vol. 1 (Princeton University Press, 1996).

3. Charles Darwin, *On the Origin of Species by Means of Natural Selection, or The Preservation of Favoured Races in the Struggle for Life* (J. Murray, 1859).

4. A. Hill, "Aspects of Genetic Susceptibility to Human Infectious Diseases," *Annual Review of Genetics* 40, no. 1 (2006): 469–486.

5. Kathleen LaRow Brown, Vijendra Ramlall, Michael Zietz, Undina Gisladottir, and Nicholas P. Tatonetti, "Estimating the Heritability of SARS-CoV-2 Susceptibility and COVID-19 Severity," *Nature Communications* 15, no. 1 (2024): 367.

6. Michelle J. K. Osterman, *Changes in Primary and Repeat Cesarean Delivery: United States 2016–2021* (Centers for Disease Control and Prevention, 2022).

7. Martin A. Schlaepfer, Michael C. Runge, and Paul W. Sherman, "Ecological and Evolutionary Traps," *Trends in Ecology & Evolution* 17, no. 10 (2002): 474–480.

8. Philipp Mitteroecker and Barbara Fischer, "Evolution of the Human Birth Canal," *American Journal of Obstetrics and Gynecology* 230, no. 3 (2024): S841–S855.

9. Daniel Lieberman, *The Story of the Human Body: Evolution, Health, and Disease* (Vintage, 2014).

10. Lance J. Kriegsfeld and Rae Silver, "The Regulation of Neuroendocrine Function: Timing Is Everything," *Hormones and Behavior* 49, no. 5 (2006): 557–574.

11. Cynthia M. Beall, "Adaptation to High Altitude: Phenotypes and Genotypes," *Annual Review of Anthropology* 43, no. 1 (2014): 251–272.

12. Kristina Beblo-Vranesevic, Harald Huber, and Petra Rettberg, "High Tolerance of *Hydrogenothermus marinus* to Sodium Perchlorate," *Frontiers in Microbiology* 8 (2017): 1369.

13. It's worth noting that cancers that affect people later in life are less subject to natural selection than cancers that affect younger people. The reason is that any gene that has an effect on an individual after they have already had kids has no effect on how many copies of their genes make it to later generations. This is thought to be one of the reasons that cancer is more common among older adults.

14. Nancy A. Moran and Tyler Jarvik, "Lateral Transfer of Genes from Fungi Underlies Carotenoid Production in Aphids," *Science* 328, no. 5978 (2010): 624–627; George Klein, "Toward a Genetics of Cancer Resistance," *Proceedings of the National Academy of Sciences* 106, no. 3 (2009): 859–863.

15. Douglas Futuyma and Mark Kirkpatrick, *Evolution*, 5th ed. (Oxford University Press, 2022).

16. Shane C. Campbell-Staton, Brian J. Arnold, Dominique Gonçalves, Petter Granli, Joyce Poole, Ryan A. Long, and Robert M. Pringle, "Ivory Poaching and the Rapid Evolution of Tusklessness in African Elephants," *Science* 374, no. 6566 (2021): 483–487.

17. See "Joining the Whitney Expedition: Ernst Mayr, Scientist," Web of Stories, https://www.webofstories.com/play/ernst.mayr/13.

18. E. Mayr, *Systematics and the Origin of Species* (Columbia University Press, 1942).

19. In the first half of the twentieth century, Darwin's theory of evolution was combined with the emerging science of genetics tied together with the quantitative rigor of statistics in what Julian Huxley dubbed "the Modern Synthesis."

20. J. Bristol Foster, "Evolution of Mammals on Islands," *Nature* 202, no. 4929 (1964): 234–235.

21. Leigh Van Valen, "Patterns and the Balance of Nature," *Evolutionary Theory* 1 (1973): 31–49; Ana Benítez-López, Luca Santini, Juan Gallego-Zamorano, Borja Milá, Patrick Walkden, Mark A. J. Huijbregts, and Joseph A. Tobias, "The Island Rule Explains Consistent Patterns of Body Size Evolution in Terrestrial Vertebrates," *Nature Ecology & Evolution* 5, no. 6 (2021): 768–786.

22. Oliver Sacks, *The Island of the Colorblind* (Vintage Books, 1996).

23. The grandmother character in the Disney movie *Moana* was based on Hinano Teavai-Murphy, who served as a cultural consultant for the film.

24. See "Taputapuātea," World Heritage Convention, UNESCO.org, https://whc.unesco.org/en/list/1529/.

25. Marshall I. Weisler, Robert Bolhar, Jinlong Ma, Emma St Pierre, Peter Sheppard, Richard K. Walter, Yuexing Feng, et al., "Cook Island Artifact Geochemistry Demonstrates Spatial and Temporal Extent of Pre-European Interarchipelago Voyaging in East Polynesia," *Proceedings of the National Academy of Sciences* 113, no. 29 (2016): 8150–8155.

26. The one exception seems to be Rapa Nui, or Easter Island, which seems to have had no contact after it was settled. But this was also one of the last islands to be settled, around 1200 CE. As such, it had perhaps the shortest amount of time after being settled until the arrival of Europeans in 1722.

27. Alexander G. Ioannidis, Javier Blanco-Portillo, Karla Sandoval, Erika Hagelberg, Carmina Barberena-Jonas, Adrian V. S. Hill, Juan Esteban Rodríguez-Rodríguez, et al., "Paths and Timings of the Peopling of Polynesia Inferred from Genomic Networks," *Nature* 597, no. 7877 (2021): 522–526.

28. Miles C. Benton, Shani Stuart, Claire Bellis, Donia Macartney-Coxson, David Eccles, Joanne E. Curran, Geoff Chambers, et al., "'Mutiny on the Bounty': The Genetic History of Norfolk Island Reveals Extreme Gender-Biased Admixture," *Investigative Genetics* 6 (2015): 1–8.

29. Stuart Macgregor, Claire Bellis, Rod A. Lea, Hannah Cox, Tom Dyer, John Blangero, Peter M. Visscher, et al., "Legacy of Mutiny on the Bounty: Founder Effect and Admixture on Norfolk Island," *European Journal of Human Genetics* 18, no. 1 (2010): 67–72.

30. Sewall Wright, *The Roles of Mutation, Inbreeding, Crossbreeding, and Selection in Evolution* (Proceedings of the VI International Congress of Genetics, 1932), 355–366.

31. Daniel Stone, "On Island of the Colorblind, Paradise Has a Different Hue," *National Geographic*, September 20, 2024, https://education.nationalgeographic.org/resource/island-colorblind-paradise-has-different-hue/.

32. Ernst Mayr, "Change of Genetic Environment and Evolution," in *Evolution as a Process*, ed. J. Huxley (Allen & Unwin, 1954).

33. Peter R. Grant and B. Rosemary Grant, *How and Why Species Multiply: The Radiation of Darwin's Finches* (Princeton University Press, 2007).

34. Patrick M. O'Grady and Rob DeSalle, "Phylogeny of the Genus *Drosophila*," *Genetics* 209, no. 1 (2018): 1–25.

35. While tortoises generally cannot swim, they can float, and instances of giant tortoises floating out at sea have been observed by mariners.

36. See "'Hobbits' on Flores, Indonesia," National Museum of Natural History, Smithsonian Institution, https://humanorigins.si.edu/research/asian-research-projects/hobbits-flores-indonesia.

37. Griffith University, "Smallest Arm Bone in the Human Fossil Record Sheds Light on the Dawn of *Homo floresiensis*," Phys.org, August 6, 2024, https://phys.org/news/2024-08-smallest-arm-bone-human-fossil.html.

38. Josh Davis, "*Homo luzonensis*: New Species of Ancient Human Discovered in the Philippines," Natural History Museum, April 11, 2019, https://www.nhm.ac.uk/discover/news/2019/april/new-species-of-ancient-human-discovered-in-the-philippines.html.

39. Cameron M. Smith, "Estimation of a Genetically Viable Population for Multigenerational Interstellar Voyaging: Review and Data for Project Hyperion," *Acta Astronautica* 97 (2014): 16–29.

40. Anders Bergström, Shane A. McCarthy, Ruoyun Hui, Mohamed A. Almarri, Qasim Ayub, Petr Danecek, Yuan Chen, et al., "Insights into Human Genetic Variation and Population History from 929 Diverse Genomes," *Science* 367, no. 6484 (2020): eaay5012.

CHAPTER 7

1. Daisy Dobrijevic, "Distance to Mars: How Far Away Is the Red Planet?" Space .com, February 4, 2022, https://www.space.com/16875-how-far-away-is-mars.html.

2. . . . other than Earth, it must be noted.

3. Robert Zubrin, *The New World on Mars: What We Can Create on the Red Planet* (Diversion Books, 2024).

4. "Can We Feed the World and Ensure No One Goes Hungry?," UN News, October 3, 2019, https://news.un.org/en/story/2019/10/1048452.

5. Julie J. Lesnik, *Edible Insects and Human Evolution* (University Press of Florida, 2019).

6. Several different sources of rock have been used as simulated Martian regolith. Volcanic ash from a cinder cone called Pu'u Nene on the lower slopes of Mauna Kea in Hawaii was used by NASA beginning in 1997 after finding that its characteristics were similar to the readings from the Viking landing sites on Mars. But experiments found that it absorbed water more readily than was expected for real Martian regolith. A mine on Saddleback Mountain in California's Mojave Desert was found to be a close match to measurements of the regolith taken on Mars by the Pathfinder and Spirit rovers. Material from this site has been widely used by researchers and is currently being sold for use in research and education (or just for fun) by a company called the Martian Garden. G. W. Wieger Wamelink, Joep Y. Frissel, Wilfred H. J. Krijnen, M. Rinie Verwoert, and Paul W. Goedhart, "Can Plants Grow on Mars and the Moon: A Growth Experiment on Mars and Moon Soil Simulants," *PLOS One* 9, no. 8 (2014): e103138.

7. A. Eichler, N. Hadland, D. Pickett, D. Masaitis, D. Handy, A. Perez, D. Batcheldor, et al., "Challenging the Agricultural Viability of Martian Regolith Simulants," *Icarus* 354 (2021): 114022.

8. Andy Weir, *The Martian: A Novel* (Broadway Books, 2016).

9. G. W. W. Wamelink, J. Y. Frissel, W. H. J. Krijnen, and M. R. Verwoert, "Crop Growth and Viability of Seeds on Mars and Moon Soil Simulants," *Open Agriculture* 4, no. 1 (2019): 509–516.

10. Luigi Giuseppe Duri, Antonio Giandonato Caporale, Youssef Rouphael, Simona Vingiani, Mario Palladino, Stefania De Pascale, and Paola Adamo, "The Potential for Lunar and Martian Regolith Simulants to Sustain Plant Growth: A Multidisciplinary Overview," *Frontiers in Astronomy and Space Sciences* 8 (2022): 747821.

11. Marcel G. A. van Der Heijden, Francis M. Martin, Marc-André Selosse, and Ian R. Sanders, "Mycorrhizal Ecology and Evolution: The Past, the Present, and the Future," *New Phytologist* 205, no. 4 (2015): 1406–1423.

12. Declan Watts, Enzo A. Palombo, Alex Jaimes Castillo, and Bita Zaferanloo, "Endophytes in Agriculture: Potential to Improve Yields and Tolerances of Agricultural Crops," *Microorganisms* 11, no. 5 (2023): 1276.

13. Randall Rainwater and Arijit Mukherjee, "The Legume-Rhizobia Symbiosis Can Be Supported on Mars Soil Simulants," *PLOS One* 16, no. 12 (2021): e0259957.

14. Ed Yong, *I Contain Multitudes: The Microbes Within Us and a Grander View of Life* (Random House, 2016).

15. Francine E. Garrett-Bakelman, Manjula Darshi, Stefan J. Green, Ruben C. Gur, Ling Lin, Brandon R. Macias, Miles J. McKenna, et al., "The NASA Twins Study: A Multidimensional Analysis of a Year-Long Human Spaceflight," *Science* 364, no. 6436 (2019): eaau8650.

16. Braden T. Tierney, JangKeun Kim, Eliah G. Overbey, Krista A. Ryon, Jonathan Foox, Maria A. Sierra, Chandrima Bhattacharya, et al., "Longitudinal Multi-omics Analysis of Host Microbiome Architecture and Immune Responses During Short-Term Spaceflight," *Nature Microbiology* (2024): 1–15.

17. Neil Armstrong celebrated his thirty-ninth birthday while under quarantine in Building 37 at Johnson Space Center. The medical team, the only other people allowed inside the facility, surprised him with a cake.

18. Gerald R. Taylor, Richard C. Graves, Royce M. Brockett, J. Kelton Ferguson, and Ben J. Mieszkuc, "Skylab Environmental and Crew Microbiology Studies," *Biomedical Results from Skylab* (NASA, 1977).

19. See "Shuttle-Mir Oral Histories," NASA.gov, https://www.nasa.gov/history/SP-4225/oral-histories/oral-histories.htm.

20. C. M. Ott, R. J. Bruce, and D. L. Pierson, "Microbial Characterization of Free Floating Condensate Aboard the Mir Space Station," *Microbial Ecology* 47 (2004): 133–136.

21. Snehit Mhatre, Jason M. Wood, Aleksandra Checinska Sielaff, Maximilian Mora, Stefanie Duller, Nitin Kumar Singh, Fathi Karouia, et al., "Assessing the Risk of Transfer of Microorganisms at the International Space Station due to Cargo Delivery by Commercial Resupply Vehicles," *Frontiers in Microbiology* 11 (2020): 566412.

22. Jenna M. Lang, David A. Coil, Russell Y. Neches, Wendy E. Brown, Darlene Cavalier, Mark Severance, Jarrad T. Hampton-Marcell, et al., "A Microbial Survey of the International Space Station (ISS)," *PeerJ* 5 (2017): e4029.

23. Rob Dunn, *Never Home Alone: From Microbes to Millipedes, Camel Crickets, and Honeybees, the Natural History of Where We Live* (Basic Books, 2018).

24. Aleksandra Checinska Sielaff, Camilla Urbaniak, Ganesh Babu Malli Mohan, Victor G. Stepanov, Quyen Tran, Jason M. Wood, Jeremiah Minich, et al., "Characterization of the Total and Viable Bacterial and Fungal Communities Associated with the International Space Station Surfaces," *Microbiome* 7 (2019): 1–21.

25. Braden T. Tierney, Nitin K. Singh, Anna C. Simpson, Andrea M. Hujer, Robert A. Bonomo, Christopher E. Mason, and Kasthuri Venkateswaran, "Multidrug-Resistant *Acinetobacter pittii* Is Adapting to and Exhibiting Potential Succession Aboard the International Space Station," *Microbiome* 10, no. 1 (2022): 210.

26. Swati Bijlani, Nitin K. Singh, V. V. Eedara, Appa Rao Podile, Christopher E. Mason, Clay C. C. Wang, and Kasthuri Venkateswaran, "*Methylobacterium ajmalii sp. nov.*, Isolated from the International Space Station," *Frontiers in Microbiology* 12 (2021): 639396.

27. Nitin K. Singh, Daniela Bezdan, Aleksandra Checinska Sielaff, Kevin Wheeler, Christopher E. Mason, and Kasthuri Venkateswaran, "Multi-Drug Resistant *Enterobacter bugandensis* Species Isolated from the International Space Station and Comparative Genomic Analyses with Human Pathogenic Strains," *BMC Microbiology* 18 (2018): 1–13.

28. Brian Crucian, Raymond Stowe, Satish Mehta, Peter Uchakin, Heather Quiriarte, Duane Pierson, and Clarence Sams, "Immune System Dysregulation Occurs During Short Duration Spaceflight on Board the Space Shuttle," *Journal of Clinical Immunology* 33 (2013): 456–465.

29. Fei Wu, Huixun Du, Eliah Overbey, JangKeun Kim, Priya Makhijani, Nicolas Martin, Chad A. Lerner, et al., "Single-Cell Analysis Identifies Conserved Features of Immune Dysfunction in Simulated Microgravity and Spaceflight," *Nature Communications* 15, no. 1 (2024): 4795.

30. Bridgette V. Rooney, Brian E. Crucian, Duane L. Pierson, Mark L. Laudenslager, and Satish K. Mehta, "Herpes Virus Reactivation in Astronauts During Spaceflight and Its Application on Earth," *Frontiers in Microbiology* 10 (2019): 432964.

31. NASA began pre-flight quarantine following Apollo 9, during which all three crew members came down with respiratory infections that caused the launch to be delayed by three days (John Uri, "55 Years Ago: Four Months Until the Moon Landing," NASA.gov, March 20, 2024, https://www.nasa.gov/history/55-years-ago-four-months-until-the-moon-landing/).

32. David Quammen, *Spillover: Animal Infections and the Next Human Pandemic* (W. W. Norton, 2012).

33. While insects like mosquitoes often serve as vectors for infectious disease, meaning that they are in effect a delivery mechanism for the pathogen, we almost never get human diseases that were once insect diseases, as frequently happens for diseases infecting birds and mammals.

34. One way around this would be to develop lab-grown meat. Several companies are currently experimenting with growing animal cells in culture in ways that would allow them to be used as food. But this would not require having live animals and as such would still prevent the spread of zoonotic diseases.

35. David P. Strachan, "Hay Fever, Hygiene, and Household Size," *BMJ: British Medical Journal* 299, no. 6710 (1989): 1259.

36. Leena Von Hertzen, Ilkka Hanski, and Tari Haahtela, "Natural Immunity: Biodiversity Loss and Inflammatory Diseases Are Two Global Megatrends That Might Be Related," *EMBO Reports* 12, no. 11 (2011): 1089–1093.

37. L. Ruokolainen, L. Paalanen, A. Karkman, T. Laatikainen, L. Von Hertzen, T. Vlasoff, O. Markelova, et al., "Significant Disparities in Allergy Prevalence and Microbiota Between the Young People in Finnish and Russian Karelia," *Clinical & Experimental Allergy* 47, no. 5 (2017): 665–674.

38. "'Bubble Boy' 40 Years Later: Look Back at Heartbreaking Case," CBS News, September 21, 2011, https://www.cbsnews.com/pictures/bubble-boy-40-years-later-look-back-at-heartbreaking-case/15/.

39. Jenny Cobb, "David Vetter Isolation Suit: The Inspirational Story of 'Bubble Boy' Has Lasting Impact," Bullock Museum, n.d., https://www.thestoryoftexas.com/discover/artifacts/david-vetter-suit-spotlight-010816.

40. Todd Ackerman, "The 'Boy in the Bubble' Who Captivated the World," Chron.com, June 3, 2016, https://www.chron.com/local/history/medical-science/article/The-boy-in-the-bubble-who-captivated-the-world-7952962.php.

41. Evelyn Jane Collen, Angad Singh Johar, João C. Teixeira, and Bastien Llamas, "The Immunogenetic Impact of European Colonization in the Americas," *Frontiers in Genetics* 13 (2022): 918227; P. M. Martin and C. Combes, "Emerging Infectious Diseases and the Depopulation of French Polynesia in the 19th Century," *Emerging Infectious Diseases* 2, no. 4 (1996): 359–361.

CHAPTER 8

1. See Mason Lab, https://www.masonlab.net/.

2. Christopher E. Mason, *The Next 500 Years: Engineering Life to Reach New Worlds* (MIT Press, 2021).

3. David A. Jackson, Robert H. Symons, and Paul Berg, "Biochemical Method for Inserting New Genetic Information into DNA of Simian Virus 40: Circular SV40 DNA Molecules Containing Lambda Phage Genes and the Galactose Operon of *Escherichia coli*," *Proceedings of the National Academy of Sciences* 69, no. 10 (1972): 2904–2909.

4. See "Recombinant DNA in the Lab," National Museum of American History, Smithsonian Institution, https://americanhistory.si.edu/collections/object-groups/birth-of-biotech/recombinant-dna-in-the-lab.

5. Alok Jha, "Glow Cat: Fluorescent Green Felines Could Help Study of HIV," *Guardian*, September 11, 2011, https://www.theguardian.com/science/2011/sep/11/genetically-modified-glowing-cats.

6. K. Ingemar Jönsson, Elke Rabbow, Ralph O. Schill, Mats Harms-Ringdahl, and Petra Rettberg, "Tardigrades Survive Exposure to Space in Low Earth Orbit," *Current Biology* 18, no. 17 (2008): R729–R731.

7. Takuma Hashimoto, Daiki D. Horikawa, Yuki Saito, Hirokazu Kuwahara, Hiroko Kozuka-Hata, Tadasu Shin-i, Yohei Minakuchi, et al. "Extremotolerant Tardigrade Genome and Improved Radiotolerance of Human Cultured Cells by Tardigrade-Unique Protein," *Nature Communications* 7, no. 1 (2016): 12808.

8. Michael Rogers, "The Pandora's Box Congress," *Rolling Stone*, June 19, 1975.

9. Ernest Beutler, "The Cline Affair," *Molecular Therapy* 4, no. 5 (2001): 396–397; Marjorie Sun, "Cline Loses Two NIH Grants: Tough Stance Meant as a Signal That Infractions Will Not Be Tolerated," *Science* 214, no. 4526 (1981): 1220–1220.

10. Biotechnology Innovation Organization, "Ashanthi DeSilva—Very First Gene Therapy Patient," YouTube, June 7, 2018, https://www.youtube.com/watch?v=IgES04-cSr8.

11. Siddhartha Mukherjee, *The Gene: An Intimate History* (Scribner, 2016).

12. Walter Isaacson, *The Code Breaker: Jennifer Doudna, Gene Editing, and the Future of the human Race* (Simon and Schuster, 2021).

13. Martin Jinek, Krzysztof Chylinski, Ines Fonfara, Michael Hauer, Jennifer A. Doudna, and Emmanuelle Charpentier, "A Programmable Dual-RNA–Guided DNA Endonuclease in Adaptive Bacterial Immunity," *Science* 337, no. 6096 (2012): 816–821.

14. Prashant Mali, Luhan Yang, Kevin M. Esvelt, John Aach, Marc Guell, James E. DiCarlo, Julie E. Norville, and George M. Church, "RNA-Guided Human Genome Engineering via Cas9," *Science* 339, no. 6121 (2013): 823–826; Le Cong, F. Ann Ran, David Cox, Shuailiang Lin, Robert Barretto, Naomi Habib, Patrick D. Hsu, et al., "Multiplex Genome Engineering Using CRISPR/Cas Systems," *Science* 339, no. 6121 (2013): 819–823; Martin Jinek, Alexandra East, Aaron Cheng, Steven Lin, Enbo Ma, and Jennifer Doudna, "RNA-Programmed Genome Editing in Human Cells," *eLife* 2 (2013): e00471.

15. Jia Yi Tan, Yong Hao Yeo, Kok Hoe Chan, Hamid Shaaban, and Gunwant Guron, "Mortality Trends and Causes of Death in Individuals with Thalassemia: A Population-Based Retrospective Study in the United States from 1999 to 2020," *Blood* 142 (2023): 3740; Yuanyuan Tuo, Yang Li, Yan Li, Jianjuan Ma, Xiaoyan Yang, Shasha Wu, Jiao Jin, and Zhixu He, "Global, Regional, and National Burden of Thalassemia, 1990–2021: A Systematic Analysis for the Global Burden of Disease Study 2021," *eClinicalMedicine* 72 (2024): 102619.

16. Steve Connor, "Nobel Scientist Happy to 'Play God' with DNA," *Independent* 17 (2000).

17. David Baltimore, Paul Berg, Michael Botchan, Dana Carroll, R. Alta Charo, George Church, Jacob E. Corn, et al., "A Prudent Path Forward for Genomic Engineering and Germline Gene Modification," *Science* 348, no. 6230 (2015): 36–38.

18. National Academies of Sciences, Engineering, and Medicine, *International Summit on Human Gene Editing: A Global Discussion* (National Academies Press, 2016).

19. David Baltimore, R. Alta Charo, Daniel J. Kevles, and Ruha Benjamin, "Summit on Human Gene Editing," *National Academies of Sciences* 32, no. 3 (2016): 61–69.

20. The He Lab, "About Lulu and Nana: Twin Girls Born Healthy After Gene Surgery as Single-Cell Embryos," YouTube, November 25, 2018, https://www.youtube.com/watch?v=th0vnOmFltc.

21. Marilynn Marchione, "Chinese Researcher Claims First Gene-Edited Babies," AP News, November 26, 2018, https://apnews.com/article/ap-top-news-international-news-ca-state-wire-genetic-frontiers-health-4997bb7aa36c45449b488e19ac83e86d; Antonio Regalado, "Exclusive: Chinese Scientists Are Creating CRISPR Babies," *MIT Technology Review*, November 25, 2018, https://www.technologyreview.com/2018/11/25/138962/exclusive-chinese-scientists-are-creating-crispr-babies/; https://apnews.com/article/13303d99c4f849829e98350301e334a9; Ian Sample, "Chinese Scientist Who Edited Babies' Genes Jailed for Three Years," *Guardian*, December 20, 2019, https://www.theguardian.com/world/2019/dec/30/gene-editing-chinese-scientist-he-jiankui-jailed-three-years.

22. Isaacson, *The Code Breaker*.

23. Mason, *The Next 500 Years*.

24. Francis Galton, *Inquiries into Human Faculty and Its Development* (Dent, 1883).

25. *Buck v. Bell*, 274 US 200 (1927).

26. Carl Zimmer, *She Has Her Mother's Laugh: The Powers, Perversions, and Potential of Heredity* (Penguin, 2019).

27. Jennifer A. Doudna and Samuel H. Sternberg, *A Crack in Creation: Gene Editing and the Unthinkable Power to Control* (Mariner Books, 2017).

28. Mason, *The Next 500 Years*.

29. Lindsay A. Rutter, Matthew J. MacKay, Henry Cope, Nathaniel J. Szewczyk, Jang-Keun Kim, Eliah Overbey, Braden T. Tierney, et al., "Protective Alleles and Precision Healthcare in Crewed Spaceflight," *Nature Communications* 15, no. 1 (2024): 6158; Consortium for Space Genetics, Harvard Medical School, https://spacegenetics.hms.harvard.edu/.

30. Michael M. Cox and John R. Battista. "*Deinococcus radiodurans*: The Consummate Survivor," *Nature Reviews Microbiology* 3, no. 11 (2005): 882–892; Edmond Jolivet, Stéphane L'Haridon, Erwan Corre, Patrick Forterre, and Daniel Prieur, "*Thermococcus gammatolerans sp. nov.*, a Hyperthermophilic Archaeon from a Deep-Sea Hydrothermal Vent That Resists Ionizing Radiation," *International Journal of Systematic and Evolutionary Microbiology* 53, no. 3 (2003): 847–851.

31. Clyde A. Hutchison III, Ray-Yuan Chuang, Vladimir N. Noskov, Nacyra Assad-Garcia, Thomas J. Deerinck, Mark H. Ellisman, John Gill, et al., "Design and Synthesis of a Minimal Bacterial Genome," *Science* 351, no. 6280 (2016): aad6253; Amor A. Menezes, Michael G. Montague, John Cumbers, John A. Hogan, and Adam P. Arkin, "Grand Challenges in Space Synthetic Biology," *Journal of the Royal Society Interface* 12, no. 113 (2015): 20150803.

32. While horses and donkeys can mate, producing hybrids known as mules, the mules themselves are sterile.

33. John Hawks, "When Did Human Chromosome 2 Fuse?," Johnhawks.net, August 31, 2023, https://johnhawks.net/weblog/when-did-human-chromosome-2-fuse/.

34. Aleksandra Kawala-Sterniuk, Natalia Browarska, Amir Al-Bakri, Mariusz Pelc, Jaroslaw Zygarlicki, Michaela Sidikova, Radek Martinek, and Edward Jacek Gorzelanczyk, "Summary of Over Fifty Years with Brain-Computer Interfaces: A Review," *Brain Sciences* 11, no. 1 (2021): 43; Richard Martin, "Mind Control," *Wired*, March 1, 2005, https://www.wired.com/2005/03/brain-3/; "'We Did It!' Brain-Controlled 'Iron Man' Suit Kicks Off World Cup," NBC News, June 12, 2014, https://www.nbcnews.com/storyline/world-cup/we-did-it-brain-controlled-iron-man-suit-kicks-world-n129941.

35. "Neuralink's First Human Patient Able to Control Mouse Through Thinking, Musk Says," Reuters, February 20, 2024, https://www.reuters.com/business/healthcare-pharmaceuticals/neuralinks-first-human-patient-able-control-mouse-through-thinking-musk-says-2024-02-20/.

36. Manfred E. Clynes and Nathan S. Kline, "Cyborgs and Space," *Astronautics* 14, no. 9 (1960): 26–27.

37. See "Bina: Custom Character Robot," Hanson Robotics, https://www.hansonrobotics.com/bina48-9/.

EPILOGUE

1. Eric Berger, *Liftoff: Elon Musk and the Desperate Early Days That Launched SpaceX* (William Morrow, an imprint of HarperCollinsPublishers, 2021).

2. See Everyday Astronaut, YouTube channel, https://www.youtube.com/c/everydayastronaut/live.

3. Everyday Astronaut, "[4K] Watch SpaceX Launch Starship, LIVE Up Close and Personal!" YouTube, March 14, 2024, https://www.youtube.com/watch?v=ixZpBOxMopc.

4. Eric Berger, "After Thursday's Flight, Starship Is Already the Most Revolutionary Rocket Ever Built," *Ars Technica*, March 15, 2024, https://arstechnica.com/space/2024/03/thursdays-starship-flight-provided-a-glimpse-into-a-future-of-abundant-access-to-space/.

5. Kelly Weinersmith and Zach Weinersmith, *A City on Mars: Can We Settle Space, Should We Settle Space, and Have We Really Thought This Through?* (Random House, 2023).

6. Savannah Mandel, *Ground Control: An Argument for the End of Human Space Exploration* (Chicago Review Press, 2024).

INDEX